Innovations in Food Processing

T0201104

FOOD PRESERVATION TECHNOLOGY SERIES

Series Editor
Gustavo V. Barbosa-Cánovas

Innovations in Food Processing
Editors: Gustavo V. Barbosa-Cánovas and Grahame W. Gould

Trends in Food Engineering
Editors: Jorge E. Lozano, Cristina Añón, Efrén Parada-Arias,
and Gustavo V. Barbosa-Cánovas

**Pulsed Electric Fields in Food Processing:
Fundamental Aspects and Applications**
Editors: Gustavo V. Barbosa-Cánovas and Q. Howard Zhang

**Osmotic Dehydration and Vacuum Impregnation:
Applications in Food Industries**
Editors: Pedro Fito, Amparo Chiralt, Jose M. Barat, Walter E. L. Spiess,
and Diana Behsnilian

Engineering and Food for the 21st Century
Editors: Jorge Welti-Chanes, Gustavo V. Barbosa-Cánovas,
and José Miguel Aguilera

Unit Operations in Food Engineering
Albert Ibarz and Gustavo V. Barbosa-Cánovas

Transport Phenomena in Food Processing
Editors: Jorge Welti-Chanes, Jorge F. Vélez-Ruiz,
and Gustavo V. Barbosa-Cánovas

FOOD PRESERVATION TECHNOLOGY SERIES

Innovations in Food Processing

EDITED BY

Gustavo V. Barbosa-Cánovas, Ph.D.
Washington State University

Grahame W. Gould, Ph.D.
Unilever Research Laboratory

CRC Press
Taylor & Francis Group
Boca Raton London New York

CRC Press is an imprint of the
Taylor & Francis Group, an **informa** business

CRC Press
Taylor & Francis Group
6000 Broken Sound Parkway NW, Suite 300
Boca Raton, FL 33487-2742

First issued in paperback 2019

ISBN-13: 978-1-56676-782-8 (hbk)
ISBN-13: 978-0-367-39851-4 (pbk)
Library of Congress Card Number 00-104017

Library of Congress Cataloging-in-Publication Data
Main entry under title: Food Preservation Technology Series: Innovations in Food Processing

Visit the Taylor & Francis Web site at
http://www.taylorandfrancis.com

and the CRC Press Web site at
http://www.crcpress.com

To our families

Table of Contents

Series Preface

THE processing of foods is becoming more sophisticated and diverse, in response to the growing demand for quality foods. Consumers today expect food products to provide, among other things, convenience, variety, adequate shelf life and caloric content, reasonable cost, and environmental soundness. Strategies to meet these demands include modifications to existing food processing techniques and the adoption of novel processing technologies.

This new Technomic Book Series has been conceived to explore the food processing techniques that will facilitate the transformation of the food market to meet consumer expectations. The Series will include titles on fundamental aspects in food processing, such as advances in transport phenomena, as well as titles covering more applied topics, such the present book, *Innovations in Food Processing*. Other books to follow will discuss trends in food engineering, advances in pulsed electric fields, alternative technologies for processing foods, the interface of food science and biotechnology, and major topics in engineering and food for the 21st century. We expect to publish two to four new titles per year and hope this Series will become a premier reference for readers seeking the latest developments in food processing.

Our challenge, which we face with enthusiasm and optimism, is to bring readers quality publications on what it is current, and what may become the food preservation technologies of the future. We hope this effort will be well received by the food industry, food related agencies, universities, and other research institutions.

GUSTAVO V. BARBOSA-CÁNOVAS
Series Editor

Preface

INNOVATION is one of the keys for the sustained growth of the food industry, even though the pathways from concept to implementation are not trivial, and many times are quite painful. One of the reasons these pathways are so bumpy is that the hurdles in the road to implementation are not properly address or understood. In order to significantly increase the chances for success, basic research covering a broad spectrum of disciplines and expectations is necessary prior to the commercialization of new products and technologies. At the same time, it is apparent that consumers all around the world are becoming more knowledgeable about food products, regulatory agencies more stringent, and the food industry more liable. Therefore, in order to come up with better quality food products, we need to make every effort to understand principles, be aware of new opportunities, and consider the combination of strategies. These days, the food world has a handful of options to explore to make the food industry more diverse, competitive and efficient. The aim of this book is to investigate some of these options, alternative technologies and strategies to properly address the new challenges facing the food industry, and to provide specific examples on how these alternatives could be utilized in some specific food products.

The first chapter of the book reviews what are the most promising emerging technologies to process foods. It is followed by eight chapters dealing with processing of foods with high hydrostatic pressure (HHP). The first in this set is a review chapter followed by research studies on the water absorption characteristics of black beans, the effect of this technology on selected pectolytic enzymes, the control of browning in apple slices, and the viscoelastic properties of egg and surimi gels induced by HHP. Chapter 10 reviews advances in minimally processed fruits using the "hurdle technology" approach, whereas the following chapter covers recent developments on minimal processing of food with thermal methods. Microwave blanching of fruits and vegetables is reviewed in chapter 12. The following chapter deals with the modeling and simulation of microbial survival in foods processed by combined methods. This technology is revisited in the next two chapters where the texture and microstructure as well as the shelf-life of apple slices are analyzed. The last three chapters are devoted to investigating the effects of polysaccharide films on apple slices, including studies on browning and quality changes during refrigerated storage.

It is very likely that most of the technologies and processing strategies presented in this book will be used by the food industry in the years to come. There is no question these innovative approaches will result in sound alternatives for addressing the always increasing demand for quality foods at a reasonable cost. While implementation at the industrial level is taking place, additional basic and applied research, as well as marketing studies, will be necessary. We sincerely hope this book will be a meaningful addition to the food literature and will promote additional interest in innovative technologies research, development and implementation.

GUSTAVO V. BARBOSA-CÁNOVAS
GRAHAME W. GOULD

Acknowledgements

THE editors want to recognize all the help received from the associate editor, Gipsy Tabilo (Washington State University), the authors, and the reviewers in making this book a reality. We also thank Jeannie Andersen and Dora Rollins, both with Washington State University, for their editorial comments.

List of Contributors

STELLA M. ALZAMORA
Departamento de Industrias
Facultad de Ciencias Exactas of the
  Universidad de Buenos Aires
Ciudad Universitaria
1428 Capital Federal, Argentina

GUSTAVO V. BARBOSA-
  CÁNOVAS
Department of Biological Systems
  Engineering
Washington State University
Pullman, WA 99164-6120 U.S.A.

NICOLÁS BRÁNCOLI
Department of Biological Systems
  Engineering
Washington State University
Pullman, WA 99164-6120 U.S.A.

TERRY BOYLSTON
Department of Food Science and
  Human Nutrition
Washington State University
Pullman, WA 99164-6376 U.S.A.

RALPH P. CAVALIERI
Department of Biological Systems
  Engineering
Washington State University
Pullman, WA 99164-6120 U.S.A.

LIDIA DORANTES-ALVAREZ
Departamento de Graduados en
  Alimentos
Escuela Nacional de Ciencias
  Biológicas
Instituto Politécnico Nacional
Carpio y Plan de Ayala, México
  City, Mexico

CARMEN GONZÁLEZ
Departament d'Enginyeria Química
Universitat de Barcelona
Martí i Franqués 1
08028 Barcelona, Spain

GRAHAME W. GOULD
17 Dove Road
Bedford MK41 7AA U.K.

GUSTAVO GUTIERREZ-LÓPEZ
Departamento de Graduados en
  Alimentos
Escuela Nacional de Ciencias
  Biológicas
Instituto Politécnico Nacional
Carpio y Plan de Ayala, México
  City, Mexico

ALBERT IBARZ
Departament de Tecnología
  d'Aliments
Universitat de Lleida
Av. Rovira Roure 177
25198 Lleida, Spain

DIETRICH KNORR
Department of Food Process
  Engineering and Food
  Biotechnology
Berlin University of Technology
Königin-Luise Str. 22
D-14195, Berlin, Germany

AURELIO LÓPEZ-MALO
Departemento de Ingeniería
  Química y Alimentos
Universidad de las Américas-Puebla
Ex-Hacienda Santa Catarina Mártir
Cholula, Puebla. C.P. 72820 México

LI MA
Department of Biological Systems
  Engineering
Washington State University
Pullman, WA 99164-6120 U.S.A.

ADELMO MONSALVE-
  GONZÁLEZ
Department of Biological Systems
  Engineering
Washington State University
Pullman, WA 99164-6120 U.S.A.

THOMAS OHLSSON
SIK, The Swedish Institute for Food
  and Biotechnology
P.O. Box 5401
S-402 29 Göteborg, Sweden

MICHA PELEG
Department of Food Science
Chenoweth Laboratory
University of Massachusetts
Amherst, MA 01003 U.S.A.

EMILY H. REN
Department of Biological Systems
  Engineering
Washington State University
Pullman, WA 99164-6120 U.S.A.

ELBA SANGRONIS
Department of Food Science and
  Human Nutrition
Washington State University
Pullman, WA 99164-6376 U.S.A.

BARRY G. SWANSON
Department of Food Science and
  Human Nutrition
Washington State University
Pullman, WA 99164-6376 U.S.A.

MARIA S. TAPIA
Instituto de Ciencia y Technolgía de
  Alimentos
Facultad de Ciencias
Universidad Central de Venezuela
P.O. Box 47097
Caracas 1041-A, Venezuela

JATUPHONG VARITH
Department of Biological Systems
  Engineering
Washington State University
Pullman, WA 99164-6120 U.S.A.

JORGE WELTI-CHANES
Departemento de Ingeniería
  Química y Alimentos
Universidad de las Américas-Puebla
Ex-Hacienda Santa Catarina Mártir
Cholula, Puebla, C.P. 72820 México

CHAPTER 1

Emerging Technologies in Food Preservation and Processing in the Last 40 Years

GRAHAME W. GOULD

CONSUMERS' desires for foods that are minimally preserved and processed are encouraging the development of new methods for the inactivation of microorganisms in foods. While the efficacy of many of these methods was demonstrated many years ago, technological advances are only now beginning to make possible their commercial exploitation.

BACKGROUND

Changes in consumers' desires in recent years have led to requirements for foods that are more convenient to store and prepare for consumption, are higher in quality and freshness, are more natural, and are nutritionally healthier than before (Figure 1.1). The reactions of food scientists and technologists to these changed requirements have included research and development into less severe or "minimal" preservation and processing methods. Many of these methods have been based on the use of existing preservation methods in new ways, particularly in new combinations (Figure 1.2). However, it is important to remember that these minimal preservation technologies generally result in a reduction in the intrinsic preservation of foods and, therefore, introduce a potential reduction in their microbiological stability and safety (AAIR, 1995). Thus, in the development of new markets, it is important that any new technologies retain or even improve upon the effectiveness of preservation and the insurance of safety that might otherwise be lost. This is particularly true for the radically new technologies that offer new ways of inactivating microorgan-

1

More convenience
        -Shelf life; storage; preparation for consumption
Higher quality
        -Flavor; texture; appearance
Fresher
More natural
Nutritionally healthier
Minimally packaged
Safer

Figure 1.1 Major trends in consumer requirements.

isms in foods that are essentially low- or nonthermal (Barbosa-Cánovas et al., 1995).

EXISTING PRESERVATION TECHNIQUES

The majority of the currently employed preservation processes and techniques act by inhibiting, either partially or completely, the growth of microorganisms in foods. Few of the available techniques act primarily by inactivation (Figure 1.3). A major trend in the application of the inhibitory techniques is to employ them in new combinations that deliver effective preservation without the extreme use of any single technique, commonly known as "hurdle technology" (Leistner, 1995). This concept was first introduced nearly 30 years ago to predict the combined effects of pH and water activity reductions on the thermal processes required to commercially sterilize foods (Braithwaite and Perigo, 1971). Soon after, Leistner introduced the hurdle concept, at first for intermediate moisture meat products (Leistner and Rodel, 1975; Leistner et al., 1981). While many of the combination techniques that have been developed or are now employed were derived empirically (Tapia de Daza et al., 1996), the availability of modern computer-based predictive models of microbial growth and survival now make the selection of useful combinations to match specific requirements increasingly easy (McMeekin et al., 1993; McClure et al., 1994).

NEW AND EMERGING TECHNOLOGIES

It is interesting that, in contrast to the existing technologies, most new and emerging food preservation technologies aim to inactive microorganisms in foods rather than inhibit them. These new technologies include some that are

Milder processing
    -Minimal over-heating; less intensive heating
Fewer additives
    -Less use of chemical preservatives
Increased use of combination or hurdle technologies
Evaluation and use of naturally occurring preservation systems
Reduction in levels of salt, fat, sugar
Reduced, environmentally friendly packaging
More attention to the elimination of food poisoning microorganisms
      from the most-often contaminated foods

Figure 1.2 Reactions of the food industry.

essentially chemically or enzymically based and some that are essentially physical (Figure 1.4). While heat remains by far the most used inactivation technique, the new technologies introduce many more possibilities and mostly represent nonthermal or mild heat alternatives to conventional heat processing.

NATURAL ADDITIVES

There is a wide range of naturally occurring antimicrobial systems that operate in animals, plants, and microorganisms in such a way as to keep healthy living tissues or communities of microorganisms free from invading bacteria, yeast, and molds (Dillon and Board, 1994; Figure 1.4). Although many of these systems have been known since the early 1900s, only a few have been specialized for use in foods (Davison and Brannen, 1993). For instance, while lysozyme has been known to be a part of the mammalian antimicrobial defense system since the pioneering work of Fleming more than 70 years ago (Fleming, 1922), it has only recently been applied for food preservation. Lysozyme has been reported to be employed at levels in excess of 100 tons per annum to prevent the "blowing" of some cheeses by lysing vegetative cells of *Clostridium tyrobutyricum* outgrowing from spores (Carminati et al., 1985; Scott et al., 1987). Activation of the lactoperoxidase system has been shown to be useful for the extension of bulk milk shelf life in those countries where early pasteurization is uncommon and refrigerated transport systems are poorly developed (Pruitt and Tenovuo, 1985), but the system has not been widely exploited.

The bacteriocin nisin, first described more than 50 years ago (Mattick and Hirsch, 1947), is increasingly used to prevent spoilage in some cheeses and in some canned foods by thermophilic spore-forming bacteria (Hurst and Hoover, 1993). More than 30 other bacteriocins have been discovered, and some are being evaluated, though few are yet used in foods (Hill, 1995).

Hundreds of herb and spice compounds have also been described and have been shown to have antimicrobial properties in laboratory studies (Nychas,

---

Techniques that inhibit the growth of microorganisms in foods
    Lowered temperature
            -Chill storage; frozen storage
    Lowered water activity
            -Drying; curing with added salt; conserving with added sugar
    Lowered pH
            -Acidification; fermentation
    Vacuum packaging
            -Removal of oxygen
    Modified atmosphere packaging
            -Addition of mixtures of CO_2; O_2; N_2
    Addition of preservatives
            -Inorganic (i.e., sulfite; nitrite)
            -Organic (i.e., sorbate; benzoate; propionate)
            -Antibiotic (i.e., nisin; natamycin)
    Control of food microstructure
            -Water-in-oil emulsions
Technologies that inactivate microorganisms in foods
    Heating
            -Pasteurization
            -Sterilization
    Restriction of microorganisms access to foods
    Packaging
    Aseptic processing

Figure 1.3 Existing technologies.

1995). While some have also been shown to be effective in foods, it has been found that their efficacy in food environments is often reduced due to binding of the compounds to food proteins or partition into fats. Furthermore, their often strong tastes and odors make them difficult to incorporate into many foods in an organoleptically acceptable manner. As a result, and because of regulatory hurdles, only a few have found a useful preservative role.

PHYSICAL TECHNIQUES

In contrast to the slow introduction of natural systems as practical food preservatives, a number of novel physical processes offer exciting alternative possibilities to heat for the inactivation of microorganisms in foods (Figure 1.4). Again, though the antimicrobial efficacies of some of these techniques have been known for many decades, the development of commercial processes has been extremely slow (Mertens and Knorr, 1992; Gould, 1995).

High Hydrostatic Pressure

Among the nonthermal physical techniques is the now well-established application of high hydrostatic pressure to inactivate vegetative microorgan-

Use of naturally occurring preservative systems
    Bacteriolytic and other enzymes
        -Lysozyme
        -Lactoperoxidase
    Non-enzymatic proteins and polypeptides
        -Nisin
        -Pediocin; other bacteriocins; culture products
        -Lactoferrin; lactoferricin
    Plant-derived antimicrobials
        -Herb and spice extracts
New and emerging physical procedures
    High hydrostatic pressure
    Combined ultrasonic, heat and pressure (manothermosonication)
    High voltage gradient pulses ("electroporation")
    Electron beam and gamma irradiation
    Laser and non-coherent light pulses
    High magnetic field pulses

Figure 1.4 New and emerging technologies.

isms in foods by pressure pasteurization. Vegetative forms of microorganisms are generally sensitive to pressures in the region of 400 to 600 MPa (4,000 to 6,000 atmospheres). This was first reported before the turn of the 20th century (Hite, 1899; Hite et al. 1914; Larson et al., 1918; Bassett and Machebouf, 1932). More recent work has provided additional detailed information (Barbosa-Cánovas, 1995; Knorr, 1995) that has confirmed the effects of pressure on vegetative cells of bacteria, yeasts, and molds, but with large species-to-species and strain-to-strain variations and some very large protective effects from the constituents of some foods (e.g., at low water activities) (Oxen and Knorr, 1993). Particularly worrying is that *Escherichia coli* 0157H7 was recently shown to be extremely pressure tolerant in some foods, for example, suffering no more than a 10^2-fold reduction in numbers in UHT milk following pressurization up to 800 MPa (Patterson et al., 1995).

High pressure has been exploited so far mainly for the preservation of foods in which spores are not a problem (i.e., foods in which the pH is too low for them to outgrow or in foods that are stored for limited periods of time at chilly temperatures). These restrictions on the use of high pressure at present result from the fact that bacterial spores are much more pressure tolerant than vegetative cells (Larson et al., 1918; Basset and Machebouf, 1932; Timson and Short, 1965). However, an observed synergy of pressure with other factors in the killing of spores, e.g., mild heating (Clouston and Wills, 1969; Sale et al., 1970), could change this situation and allow pressure, in combination with other factors, to be used analogously with heat sterilization to deliver long, safe shelf stability to high water activity, low acid foods. The synergy results from the fact that pressure apparently causes spores to germinate (Gould and

Sale, 1970), and the germinated forms are then sensitive to the killing effects of pressure or heat if the temperature is sufficiently high. Other synergistic effects of bacteriocins with pressure have been seen as well (Kalchayanand et al., 1994).

Manothermosonication

Ultrasonication at high enough intensities has been known for nearly 70 years to inactive vegetative bacteria (Harvey and Loomis, 1929) and has been reported to reduce the heat resistance of bacterial spores. The effect is synergistic with raised temperatures (Burgos et al., 1972), but, as the temperature is raised, the overall synergism becomes reduced. It is thought that this occurs because, as the vapor pressure of water rises, it has the effect of reducing the effectiveness of cavitation, which is the main cause of microbial death. However, application of a slight overpressure (i.e., a few bars) prevents this fall in effectiveness so that the synergism is maintained at the higher temperature. The combination procedure of manothermosonication is claimed to have potential for reducing pasteurization and sterilization temperatures, at least for pumpable liquids and maybe for semisolid foods (Earnshaw et al., 1995; Sala et al., 1995).

High-Voltage Electric Pulses

A method for inactivating microorganisms with electric fields was first patented nearly 40 years ago by Doevenspeck (1960). High-voltage electric shocks (electroporation) are most effective for the inactivation of vegetative bacteria, yeasts, and molds, while bacterial spores are much more tolerant (Qin et al., 1996). The cell membrane is one of the most important structures controlling many of the vegetative cell's homeostatic mechanism, including the maintenance of cytoplasmic pH value, the pH gradient across the membrane, and the osmotic balance of the cell. It is, therefore, not surprising that electroporation, which was shown 30 years ago to breach this structure (Hamilton and Sale, 1967), has such a lethal effect on vegetative cells (Tsong, 1991). However, the kinetics of inactivation may present problems. Nearly straight lines are seen when logs of the number of survivors are plotted against logs of the treatment times (or the number of pulses) rather than against the treatment time itself, as commonly plotted for heat (Zhang et al., 1995). Consequently, the necessary treatment time or number of pulses increases greatly if high log reductions are required. Although the reason for the resistance of spores is not known for certain, it probably derives from the inbuilt homeostatic mechanism that confers heat tolerance on spores by maintaining a relatively dehydrated cytoplasm, and this also results in a cytoplasm that is relatively electrically non-conducting. This would interfere with the delivery of a sufficiently high enough potential difference across the spore cytoplasmic membrane to cause damage (Hamilton and Sale, 1967). Again, synergism with other preservation factors may improve efficacy, and more recent work has already shown this. For instance, nisin and other bacteriocins have recently been documented to potentiate the lethal effects of electroporation (Kalchayanand et al., 1994).

Ionizing Radiation

The use of ionizing irradiation at pasteurization doses is legal in more than 30 countries now (Loaharanu, 1995), having first been patented for food preservation more than 60 years ago (Wurst, 1930). It is technologically relatively simple to apply, with straightforward inactivation kinetics and geometry that makes dose control and processing requirements much easier than for many heat processes. The potential values to consumers in the area of prevention of food poisoning through the elimination of pathogens, such as *salmonellae* and *campylobacters*, from some foods of animal origin and some seafoods is substantial. However, this is not widely recognized by consumers, so that slow acceptance by the public continues to hamper its widespread introduction in many countries of the world.

High-Intensity Light Pulses

Low-power lasers were revealed to inactivate vegetative bacteria more than three decades ago (Klein et al., 1965). Later studies showed that the microbicidal effects of high-intensity laser and non-coherent light pulses offer new practical approaches to the decontamination of food and packaging material surfaces, and possibly transparent foods also (Dunn et al., 1988). While the killing effects may sometimes result partially from local heating, physical removal of microorganisms from surfaces (ablation), and ultraviolet irradiation, additional non-UV effects have been claimed and form the basis for some detailed research into mechanisms (Ward et al., 1996).

High-Intensity Magnetic Fields

More rece :y, oscillating magnetic fields have been claimed to have a variety of effects on biological tissues, ranging from selective inactivation of malignant cells with little damage to normal tissues (Costa and Hoffman, 1987), to the inactivation of microorganisms on packaging materials (Hoffman, 1985). It has been proposed that the magnetic energy couples into dipolar molecules and may change membrane fluidity and interfere with ion fluxes across cell membranes (Pothakamury et al., 1993). However, inactivation of bacterial spores has been not reported, and efficacies of the treatments have not exceeded about 100-fold reductions in the number of vegetative bacteria (i.e., *Streptococcus thermophilus* inoculated into milk; *Saccharomyces cerevisiae* in orange juice; mold spores in bread) (Hoffman, 1985), so practical applications for the technique seem to be limited (Mertens and Knorr, 1992; Barbosa-Cánovas et al., 1995).

MICROBIOLOGICAL SAFETY

One aspect of the new technologies that is sometimes overlooked, especially by those engaged in initial laboratory studies, is the question of safety. Particularly for those techniques that aim to inactivate microorganisms in foods, there must be some basis for judging the efficacy required to deliver a sufficiently microbiologically safe product. For thermal processing, this has traditionally been based on the particular inactivation kinetics typically seen when microorganisms are heated. For the sterilization of high water activity and low acid foods, the 12-D concept of targeting spores of proteolytic strains of *Clostridium botulinum* has remained the theoretical basis of safe processing for many years and has consistently been proved to be very safe. For pasteurization, 6-D or 8-D processes are often concluded to be suitable targets for the destruction of vegetative pathogens.

When considering radically new processes, however, it is arguably insufficient to simply try to adapt thermal processing rationales. For instance, the kinetics of inactivation may be quite different (e.g., large "shoulders" occur on some irradiation survival curves; large "tails" have been reported on some high-pressure survival curves; the kinetics of cell death caused by high-voltage discharges are different to those of heat, etc.). In addition, some vegetative bacteria that are easily controlled by heat pasteurization may be surprisingly tolerant to alternative nonthermal pasteurization procedures (e.g., *E. coli* 0157 H7 tolerance to high pressure). It may be that some organism other than *C. botulinum* would be a preferred marker for the safe processing of high pH-high water activity foods by some of the nonthermal procedures. A substantial amount of new research and input from regulatory authorities will thus be essential as these processes are further developed.

PROS, CONS, AND RELEVANT ISSUES

There is undoubtedly a real need for new and improved preservation processes and technologies that effectively inactivate microorganisms in foods, yet maintain or improve quality and meet the other changing requirements of consumers (Figure 1.1). Such technologies are beginning to introduce, at first in niche markets, new and attractive food marketing opportunities. If the resistance of bacterial spores to some of the new technologies can be overcome in a manner that is widely proven and accepted to be safe, these markets could become immeasurably larger.

A particular attraction of techniques that act by inactivation rather than inhibition is that they potentially lead to the ultimate convenience of high-quality foods that are fully ambient stable. A further attraction is that, with regard to reducing the incidence of food poisoning, the introduction of new inactivation techniques that lead to the elimination of pathogens must be the ultimate target of primary food producers, processors, distributors, and retailers. It is important to note that the lapses of hygiene that will always occur in the food service establishment and home would be of no public health consequence if the organisms of concern did not enter these premises in the first place.

Considering the fact that the basics of most of these new technologies were known more than 60 years ago, a pertinent question to ask is why it has taken so long for the first few of them to find commercial exploitation. In most cases, the major problem has been that the status of the relevant technology was initially inadequate to support confident scale-up and economic commercialization. But now, technology is catching up.

Regarding the ultimate efficacy of new nonthermal technologies, synergism with other techniques, including traditional preservation procedures, is already being reported. Experience with other combination or hurdle approaches would indicate that future developments in this direction should further widen and improve the applicability of these techniques.

REFERENCES

AAIR Project. 1995. Physiology of food poisoning microorganisms. Special Issue of *Int. J. Food Microbiol.* 28:121–332.

Barbosa-Cánovas, G. V., Pothakamury, U. R., and Swanson, B. G. 1995. State of the art technologies for the stabilization of foods by nonthermal processes: Physical methods. In *Food Preservation by Moisture Control.* G. V. Barbosa-Cánovas and G. Welti-Chanes, eds. pp. 493–532. Lancaster, PA: Technomic Publishing Co.

Bassett, J., and Machebouf, M. A. 1932. Study on the biological effects of high pressures: Resistance of bacteria, enzymes and toxins at very high pressures. *Compt. Rendu. Hebd. Sc. Acad. Sci.* 164:1431–1442.

Braithwaite, P. J., and Perigo, J. A. 1971. The influence of pH, water activity and recovery temperature on the heat resistance and outgrowth of *Bacillus* spores. In *Spore Research.* A. N. Barker, G. W. Gould, and J. Wolf, eds. pp. 289–302. London Academic Press.

Burgos, J., Ordoñez, J. A., and Sala, F. J. 1972. Effect of ultrasonic waves on the heat resistance of *Bacillus cereus* and *Bacillus coagulans* spores. *Appl. Microbiol.* 24:497–498.

Carminati, D., Nevianti, E., and Muchetti, G. 1985. Activity of lysozyme on vegetative cells of *Clostridium tyrobutyricum. Latte.* 10:194–198.

Clouston, J. G., and Wills, P. A. 1969. Initiation of germination and inactivation of *Bacillus pumilus* spores by hydrostatic pressure. *J. Bacteriol.* 97:684–690.

Costa, J. L., and Hoffman, G. A. 1987. Malignancy treatment. US Patent 4,665,898.

Davidson, M. P., and Brannen, A. L. 1993. *Antimicrobials in Foods.* New York: Marcel Dekker Inc.

Dillon, V. M., and Board, R. G. 1994. *Natural Antimicrobial Systems and Food Preservation.* Wallingford, Oxon: CAB International.

Doevenspeck, H. 1960. German Patent 1,237,541.

Dunn, J. E., Clark, R. W., Asmus, J. F., Pearlman, J. S., Boyer, K., and Painchaud, F. 1988. Methods and apparatus for preservation of foodstuffs. Int. Patent WO 88/03369.

Earnshaw, R. G., Appleyard, J., and Hurst, R. M. 1995. Understanding physical inactivation processes: combined preservation opportunities using heat, ultrasound and pressure. *Int. J. Food Microbiol.* 28:197–219.

Fleming, A. 1922. On a remarkable bacteriolytic element found in tissues and secretions. *Proc. Roy. Soc.* B 93, 306–317.

Gould, G. W. 1995. *New Methods of Food Preservation.* Glasgow: Blackie Academic and Professional.

Gould, G. W., and Sale, A. J. H. 1970. Initiation of germination of bacterial spores by hydrostatic pressure. *J. Gen. Microbiol.* 60:335–346.

Hamilton, W. A., and Sale, A. J. H. 1967. Effects of high electric fields on microorganisms II: Mechanism of action of the lethal effect. *Biochim. Biophys. Acta.* 148:789–796.

Harvey, E., and Loomis, A. 1929. The destruction of luminous bacteria by high frequency sound waves. *J. Bacteriol.* 17:373–379.

Hill, C. 1995. Bacteriocins: Natural antimicrobials from microorganisms. Ch. 2 in *New Methods of Food Preservation.* G. W. Gould, ed. pp. 22–30. Glasgow: Blackie Academic and Professional.

Hite, B. H. 1899. The effect of pressure in the preservation of milk. *Bull. W. Virginia Univ. Expt. Stn. No.* 146:15–35.

Hite, B. H., Giddings, N. J., and Weakley, C. W. 1914. The effect of pressure on certain microorganisms encountered in the preservation of fruits and vegetables. *Bull. W. Virginia Univ. Expt. Stn. No.* 146:3–67.

Hoffman, G. A. 1985. Deactivation of microorganisms by an oscillating magnetic field. US Patent 4,524,079.

Hurst, A., and Hoover, D. G. 1993. Nisin. Ch. 10 in *Antimicrobials in Foods.* P. M. Davidson and A. C. Brannen, eds. pp. 169–394. New York: Marcel Dekker.

Kalchayanand, N., Sikes, T., Dunne, C. P., and Ray, B. 1994. Hydrostatic pressure and electroporation have increased bactericidal efficiency in combination with bacteriocins. *Appl. Environ. Microbiol.* 60:4174–4177.

Klein, E., Fine, S., Ambrus, J., Cohen, E., Neter, E., Ambrus, C., Bardos, T., and Lyman, R. 1965. Interaction of laser light with biological systems. III: Studies on biological systems *in vitro. Fed. Proc.* 24 (Suppl. 14): S104–S110.

Knorr, D. 1995. Hydrostatic pressure treatment of food: Microbiology. Ch. 8 in *New Methods of Food Preservation.* G. W. Gould, ed. pp. 159–175. Glasgow: Blackie Academic and Professional.

Larson, W. P., Hartzell, T. B., and Diehl, H. S. 1918. The effect of high pressure on bacteria. *J. Infect. Dis.* 22:271–282.

Leistner, L. 1995. Use of hurdle technology in food: Recent advances. In *Food Preservation by Moisture Control: Fundamentals and Applications.* G. V. Barbosa-Cánovas and G. Welti-Chanes, eds. pp. 377–396. Lancaster, PA: Technomic Publishing Co., Inc.

Leistner, L., and Rodel, W. 1975. The significance of water activity for microorganisms in meats. In *Water Relations of Foods.* R. B. Duckworth, ed. pp. 309–323. New York: Academic Press.

Leistner, L., Rodel, W., and Krispien, K. 1981. Microbiology of meat and meat products in high- and intermediate-moisture ranges. In *Water Activity: Influences on Food Quality*. L. B. Rockland and G. F. Stewart, eds. pp. 855–916. New York: Academic Press.

Loaharanu, P. 1995. Food irradiation: Current status and future prospects. Ch 5 in *New Methods of Food Preservation*. G. W. Gould, ed. pp. 90–111. Glasgow: Blackie Academic and Professional.

Mattick, A. T. R., and Hirsch, A. 1947. Further observations on an inhibitor (nisin) from *lactic streptococci*. *Lancet*. 2:5–6.

McClure, P. J., Blackburn, C. de W., Cole, M. B., Curtis, P. S., Jones, J. E., Legan, J. D., Ogden, I. D., Peck, M. W., Roberts, T. A., Sutherland J. P., and Walker, S. J. 1994. Modeling the growth, survival and death of microorganisms in foods: The UK micromodel approach. *Int. J. Food Microbiol*. 23:265–275.

McMeekin, T. A., Olley, J., Ross, T., and Ratkowsky, D. A. 1993. *Predictive Microbiology: Theory and Application*. Taunton: Research Studies Press Ltd.

Mertens, B., and Knorr, D. 1992. Developments of nonthermal processes for food preservation. *Food Technol*. May: 124–133.

Nychas, G. J. E. 1995. Natural antimicrobials from plants. Ch. 4 in *New Methods of Food Preservation*. G. W. Gould, ed. pp. 58–89. Glasgow: Blackie Academic and Professional.

Oxen, P., and Knorr, D. 1993. Baroprotective effects of high solute concentrations against inactivation of *Rhodotorula rubra*. *Lebensm. Wiss. Technol*. 26:220–223.

Patterson, M. F., Quinn, M., Simpson, R., and Gilmour, A. 1995. Sensitivity of vegetative pathogens to high hydrostatic pressure treatment in phosphate-buffered saline and foods. *J. Food Protect*. 58:524–529.

Pothakamury, U. R., Monsalve-González, A., Barbosa-Cánovas, G. V., and Swanson, B. G. 1993. Magnetic-field inactivation of microorganisms and generation of biological changes. *Food Technol*. 47(12):85–92.

Pruitt, K. M., and Tenovuo, J. O. 1985. *The Lactoperoxidase System: Chemistry and Biological Significance*. New York: Marcel Dekker.

Qin, B-L., Pothakamury, U. R., Barbosa-Cánovas, G. V., and Swanson, B. G. 1996. Nonthermal pasteurization of liquid foods using high-intensity pulsed electric fields. *Crit. Revs Food Sci. Nutr*. 36:603–627.

Sala, F. J., Burgos, J., Condon, S., Lopez, P., and Raso, J. 1995. Effect of heat and ultrasound on microorganisms and enzymes. Ch. 9 in *New Methods of Food Preservation*. G. W. Gould, ed. pp. 176–204. Glasgow: Blackie Academic & Professional.

Sale, A. J. H., Gould, G. W., and Hamilton, W. A. 1970. Inactivation of bacterial spores by hydrostatic pressure. *J. Gen. Microbiol*. 60:323–334.

Scott, D., Hammer, F. E., and Szalkucki, T. J. 1987. Bioconversions: Enzyme technology. In *Food Biotechnology*. D. Knorr, ed. pp. 413–442. New York: Marcel Dekker.

Tapia de Daza, M. S., Alzamora, S. M., and Welti-Chanes, G. 1996. Combination of preservation factors applied to minimal processing of foods. *Crit. Revs Food Sci. Nutr*. 36:629–659.

Timson, W. J., and Short, A. J. 1965. Resistance of microorganisms to hydrostatic pressure. *Biotechnol. Bioeng*. 7:139–159.

Tsong, T. Y. 1991. Minireview: Electroporation of cell membranes. *Biophys. J*. 60:297–302.

Ward, G. D., Watson, I. A., Stewart-Tull, D. E. S., Wardlaw, A. C., and Chatwin, C. R. 1996. Inactivation of bacteria and yeasts on agar surfaces with high power Nd: YAG laser light. *Lett. Appl. Microbiol*. 23:136–140.

Wurst, O. 1930. A method for preserving food. French Patent 701302.

Zhang, Q., Qin, B. L., Barbosa-Cánovas, G. V., and Swanson, B. G. 1995. Inactivation of *E. coli* for food pasteurization by high-strength pulsed electric fields. *J. Food Proc. Pres*. 19:103–118.

CHAPTER 2

Process Aspects of High-Pressure Treatment of Food Systems

DIETRICH KNORR

FEATURES

SUBSTANTIAL progress has been made toward the advancement of under-standing the mechanisms and kinetics of high-pressure treatment of foods and food components during the past decade. This has resulted in exciting process and product developments and has led to new commercial applications in Japan, the United States, and Europe. Extensive and multinational research activities regarding high-pressure kinetics and high-pressure applications are currently underway in Europe (with funding provided by the European Union) that concentrate on process optimization, minimal processing, and advanced and emerging technologies. High pressure is a prime example of an advanced technology that is concurrently enabling minimal ("schonend") processing to continue to grow and contribute to providing food products of the highest safety, quality, and functionality.

INTRODUCTION

The title page of the April 1997 issue of *New Scientist* reads "Is this the Future of Food?" and the heading for the cover story follows with "Squeezing the Death Out of Food. It's Delicious, Nutritional, Colorful and Above All Safe" (Hill, 1997). Basically, these statements summarize—in a journalistic nutshell—the key features, advantages, and potentials of high hydrostatic pressure processing and modification of food systems and components. Recent and key reviews on the subject include Anon (1993), Balny et al. (1992),

13

Barbosa-Cánovas and Welti-Chanes (1995), Cheftel (1995), Hayashi (1993, 1994), Hayashi and Balny (1997), Knorr (1995), Messens et al. (1997), and Tausher (1995). Most of the early work on high hydrostatic pressure was re-initiated at the University of Delaware in 1982 (Knorr, 1995) after Hite's original activities and that of Sale et al., 1970; subsequent work concentrated on the use of high hydrostatic pressure for the reduction of microorganisms (Gould, 1995; Cheftel, 1995). Only recently has there been an additional shift toward the intelligent use of the key advantages of high pressure (a waste-free technology offering a nonthermal process that affects only non-covalent bonds, enabling phase transitions, permeabilizing of biological membranes, denaturing of proteins, gelatinizing of proteins and starches, increasing reaction rates, and compacting of materials) and its consequent application for the modification of foods and food components (Weemaes et al., 1996; Messens et al., 1997), as well as its exciting potential for process improvement and development (Kalichevsky et al., 1995; Koch et al., 1996; Angersbach et al., 1997; Eshtiaghi and Knorr, 1993) and product development (Ohshima et al., 1993; Stute et al., 1996; Eshtiaghi and Knorr, 1996). This seems essential because mimicking of existing processes with new technologies has so far failed in most cases.

This chapter will concentrate on key areas of current concern including advancements in high-pressure preservation, the evaluation of shelf-life stability and quality of foods after high-pressure treatment, and the use of cultured plant cells as model systems for the evaluation of the immediate effects of high-pressure treatment on plant food.

HIGH PRESSURE FOR FOOD PRESERVATION

Although the mechanisms involved in the inactivation of microorganisms by high pressure are still not fully understood, the current accumulation of data on the death kinetics of vegetative organisms subjected to various pressure/time/temperature regimes is an essential step toward clarifying the remaining questions and aiding regulatory approval. Results of such recent activities are exemplified in Figure 2.1, where the interplay of all three variables (p, t, T) is demonstrated and the significant effect of high pressure is illustrated.

Based on these data, a three-step inactivation model has been developed where (under isobaric conditions) organisms are converted from a stable state (A) into an unstable one (B). The model takes into account the physiological response of the organisms to the stress provided by the pressure treatment (i.e., membrane permeabilization, pH shift, and protein denaturation). It has been demonstrated (Heinz and Knorr, 1996) that the organisms at stable B can recover under optimum growth conditions at atmospheric pressure. When

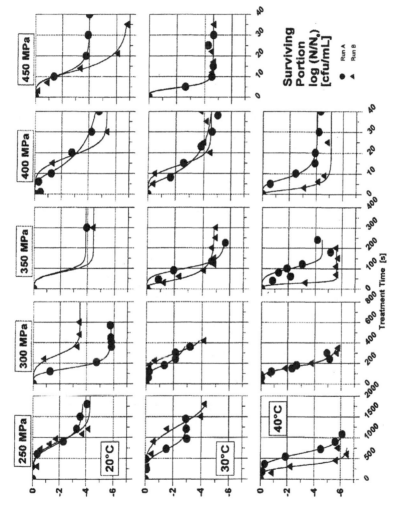

Figure 2.1 Effects of pressure, time, and temperature on the inactivation of vegetative *Bacillus subtilis* cells.

15

pressure is maintained, inactivation of *B. subtilis* takes place following first-order reaction kinetics (Figure 2.2).

Bacterial spores are considerably more resistant against high hydrostatic pressure than vegetative or germinating cells (Cheftel, 1995; Knorr, 1995). It has also been shown that pressure- or temperature-induced germination of spores is possible (Sale et al., 1970).

Based on the kinetic data accumulated in our laboratory regarding the combined action of pressure and temperature on bacterial spores, a model has been proposed (Heinz, 1997) that assumes enzyme activities of the germination pathway as the key element. Similar to alanine-induced germination (Figure 2.3; modified from Johnstone, 1994), an activated proteolytic enzyme "R" cleaves a cortex-associated "germination specific lytic enzyme" (GSLE), resulting in a successive degradation of the spore cortex peptidoglycan and a loss of typical dormant spore properties. There is increasing evidence that the unusual pressure-temperature behavior of germination kinetics is due to the complex interaction of pressure-temperature activated and/or inactivated enzyme systems involved in triggering the irreversible germination pathway. Instead of allosteric activation as the first step during the Foster-Johnstone-Pathway, a pressure-temperature effect is assumed to initialize the germination. This is consistent with the observation of accelerated enzymatic reactions under pressures not higher than 300 Mpa (Mozhaev et al., 1996).

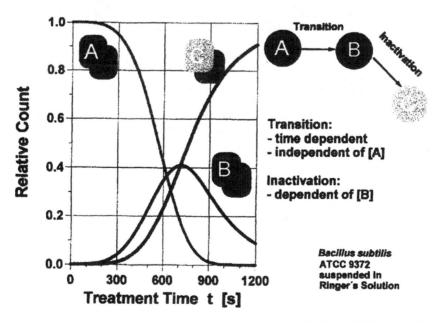

Figure 2.2 Three-state model for the inactivation of vegetative *Bacillus subtilis* cells under isobaric conditions.

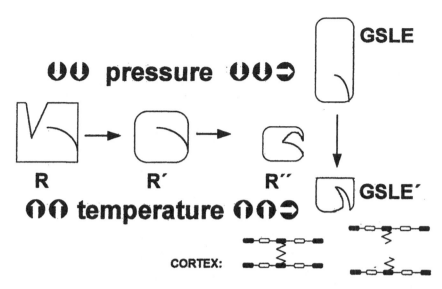

Figure 2.3 Suggested mechanisms for pressure-induced germination of bacterial spores (adapted from Johnstone, 1994).

Besides other indicators (dipicolinic acid, heat-resistance, phase-contrast microscopy), the *in vivo* measured change in optical density during pressure treatments shows clear optimum germination time when plotted versus pressure (Figure 2.4). Increasing temperature speeds up the germination rate of *B. subtilis* by accelerating the mass transfer from the core of the spore into the medium. At 38°C, no further acceleration is observed, which suggests that an effective transport limitation is stabilized in the surrounding layers of the core. It is assumed that vital parts of the germination systems are affected by the combined action of temperature and pressure. Reduced or inhibited cortex degradation may result in delayed germination or a complete loss in viability.

Despite all the recent achievements, including the work on pressure effects on cell wall ATPase (Smelt et al., 1994), as well as the effects of high pressure on the morphology of cells (Knorr, 1995), limited data are available on the interaction between pressure effects and food constituents (i.e., baroprotective effects of ingredients) and storage-time-related changes of microbial flora after pressure treatment. Cell injury and recovery—or, as Chlopin and Tamman (1903) put it, under high pressure, the majority of microorganisms tested convert in a fanlike condition from which they awake fully only after an extended period of time—require attention.

QUALITY CHANGES AFTER PRESSURE TREATMENT

Results of preliminary studies dealing with high-pressure effects on soy

Pressure Induced Spore Germination

Bacillus subtilis ATCC 9372

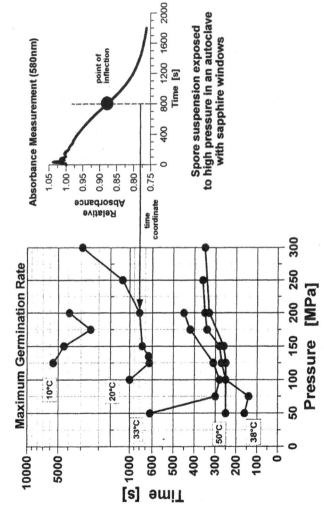

Figure 2.4 Pressure/temperature/time-induced germination of bacterial spores reached by changing the optical density of the suspension.

protein quality indicated no differences in the pattern of bands between untreated and pressure-treated samples (Figure 2.5).

Sample
1, 5, 10: protein marker
2, 9   : aqueous soy protein solution (SPL)
2   : SPL 2 h at 100°C
4   : SPL 30 min at 100°C
6   : SPL 30 min microwave heating
7   : SPL high pressure treated (600 Mpa, 20 min, 30°C)
8   : SPL high electric field pulse treated (10 kV cm^{-1}, 50 pulses)

The impact of various high-pressure treatments, including pressure shift
freezing (Kalichevsky et al., 1995) on mass and heat transfer of treated products, is demonstrated in Figures 2.6 and 2.7.

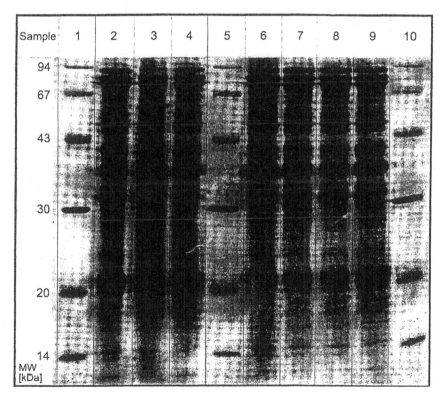

Figure 2.5 Effects of various treatments on the quality (pherogram with DISK-SDS-PAGE) of
aqueous soy protein solutions.

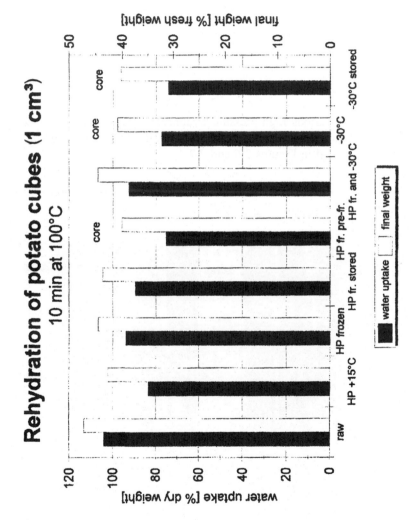

Figure 2.6 Rehydration behavior (10 min at 100°C) of differently treated potato cubes (1 cm³) followed by fluidized bed drying (70°C, 4.6 m s⁻¹).

20

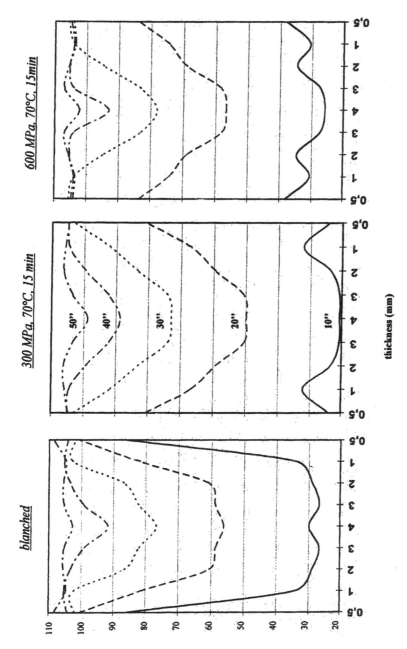

Figure 2.7 Temperature distributions within french fries (0.8 × 0.8 cm) after hot water blanching (2 min, 100°C) or pressure treatment at 300 or 600 MPa at frying (18.5°C) times between 10 and 50 seconds.

21

Raw:	untreated
HP + 15°C:	high pressure (HP), 400 Mpa, +15°C, 20 min
HP frozen:	HP frozen, 400 Mpa until −15°C, 20 min
HP fr. stored:	HP frozen, 400 Mpa until −15°C, 20 min + 10 days stored at −18°C
HP fr. pre-fr.:	1 hour frozen at −30°C, then HP frozen, 400 Mpa until −15°C, 20 min
HP fr. and −30°C:	HP frozen, 400 Mpa until −15°C, 20 min then 1 hour at −30°C
−30°C:	−30°C for 1 day
−30°C stored:	10 days stored at −18°C

It is worthwhile to note that high-pressure frozen potato cubes (Figure 2.6) resulted in the highest water uptakes during rehydration (as compared to the raw products) and that the conventionally frozen resulted in the lowest.

Temperature distributions within french fries at various frying times (10 to 50 s) clearly demonstrate how pressure treatments at various pressures heat transfer in food products, as compared to hot water blanching (Figure 2.7).

Enzyme activities constitute an integral part of the quality, preservation, and shelf-life characteristics of many food systems. A comparison of high-pressure treatments and hot water blanching on lipid oxidase and peroxidase activity in broccoli revealed more effective reduction of lipid oxidase and less effective reduction of peroxidase activity by high-pressure treatment (Figure 2.8).

It could also be shown that lipid oxidase activity [Figure 2.9(a)] was reduced more effectively after high-pressure treatment and subsequent refrigerated storage than in the water blanched samples and that it was comparable to blanching for peroxidase [Figure 2.9(b)].

Within this context, it is also interesting to note that degradation of chlorophyll within broccoli immediately after pressure treatment was found less pronounced for pressure-treated samples as compared to the blanched ones and that a temperature effect within pressure and a carbon dioxide atmosphere had a beneficial effect on reducing chlorophyll in broccoli during processing (Figure 2.10).

Raw:	untreated
Blanch:	hot water blanched, 100°C, 2 min
UHP 5°C:	high-pressure (HP) treatment (600 Mpa, 20 min) at 5°C
UHP:	HP at 20°C
UHP:	HP at 20°C, pretreated with CO_2
UHP:	HP at 50°C
Control 50°C:	50°C

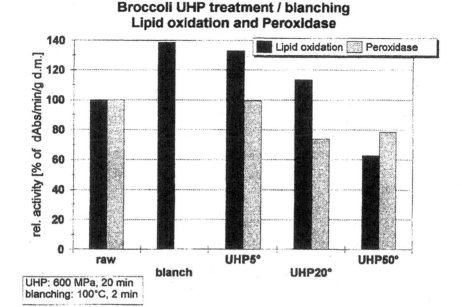

Figure 2.8 Comparison of hot water blanching and high-pressure treatment on lypoxygenase and peroxidase activity in broccoli.

These data indicate both the potential and challenge of high-pressure treatment regarding food systems and also suggest that high-pressure kinetics of quality factors differ from those of conventional thermal processing. Taking advantage of these differences and optimizing their consequences must be a major goal of future research and development activities in high-pressure treatment and processing of foods and food constituents.

PRESSURE RESPONSES OF CULTURED PLANT CELLS AS MODEL FOOD SYSTEMS

The complexity of the impact of pressure on cultured plant cells that have a great potential to serve as model systems for plant foods (Knorr, 1994) is illustrated in Figure 2.11. It appears most impressive that plant cell membranes can withstand hydrostatic pressures experienced at the bottom of the ocean (maximum approx. 100 Mpa) without major losses in cell viability (Figure 2.11). Earlier studies conducted in our laboratory illustrated that the tonoplast is more sensitive to high pressure than the outer cell membrane (Dörnenburg and Knorr, 1993). It is also noteworthy that polyphenol oxidase activity—

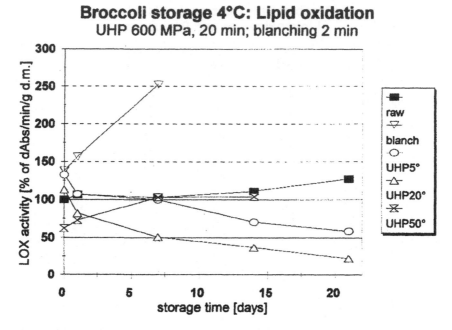

Figure 2.9 Effects of hot water blanching or high-pressure treatment on lipid oxidase (a) and peroxidase activity (b) of broccoli during 21 days at refrigerated storage.

commonly considered as being responsible for the enzymatic browning of plant foods—does not increase dramatically after pressure treatment (Figure 2.11).

However, pressure-dependent treatments of cultured potato cells at a pressure of 400 Mpa, which is high enough to permeabilize the outer cell membrane (Dörnenburg and Knorr, 1993), revealed a significant time-dependent increase in the browning of potato cells. Pressures between 0.1 and 200 Mpa affecting only the tonoplast did not seem to influence browning developments (Wille and Knorr, 1999).

This apparent contradiction could be explained via a pressure-induced increase in phenylalanine ammonia lyase activity, which in turn leads to increased phenol production. In addition, it should be taken into account that, under elevated pressures, reaction rates increase as does the interaction between enzyme and substrate (Figure 2.12). From the resulting higher polyphenol production rates, we also assume that glycosidic bonds of the phenolic substrates are affected by high-pressure treatment, which may be an additional factor in the increased browning (Dörnenburg and Knorr, 1997).

Finally, it is important to note that, based on frequency-dependent changes in specific conductivities of food systems and food constituents such as microorganisms, a method has been developed to estimate the degree of permeabili-

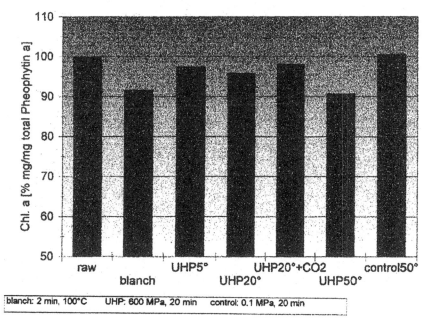

Figure 2.10 Chlorophyll degradation in broccoli immediately after different treatments.

zation a food system undergoes during various treatments (Angersbach and Knorr, 1997). The frequency-dependent charge in conductivity of potato cells treated at different pressures is illustrated in Figure 2.13. It is important to indicate that cell membrane permeabilization—and, consequently, stress and wound reactions that need to be minimized in optimized "schonend" processing operations—are less severe after pressure treatments compared to heating, freezing, or thawing (Angersbach and Knorr, 1997). This once again stresses the potential of high pressure for an almost invisible process (Labuza, 1994).

CONCLUSIONS

It is hoped that this sketchy and highly selective presentation of data from our laboratory provides a glimpse toward the potential high hydrostatic pressure treatment can offer for food research, as well as process and product design and development. Most importantly, high-pressure treatment of foods as the "third dimension" (besides time and temperature treatments of conventional

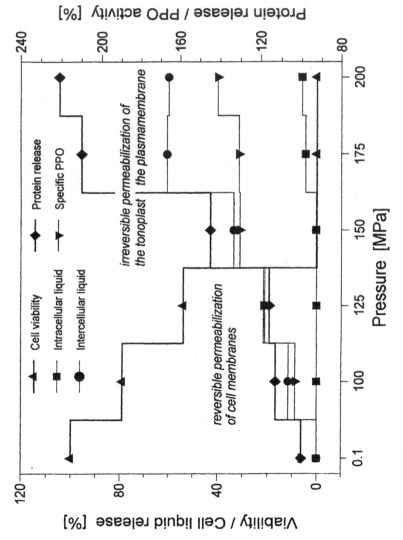

Figure 2.11 Effects of high hydrostatic pressure on the viability, cell liquid release, protein release, and polyphenol oxidase activity in a culture medium of potato cells (*Solanum tuberosum*).

26

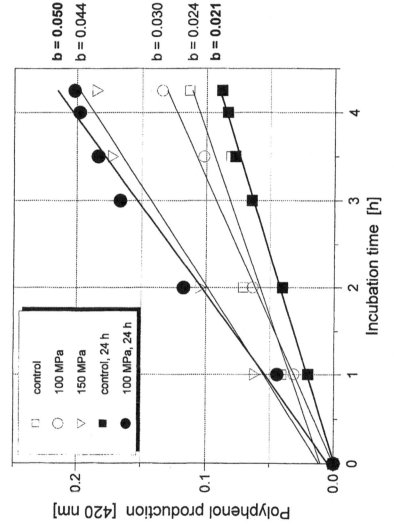

Figure 2.12 Pressure (0.1 to 150 MPa) and incubation time (0 to 4.25 h) dependent polyphenol production within cell extracts of potato cells.

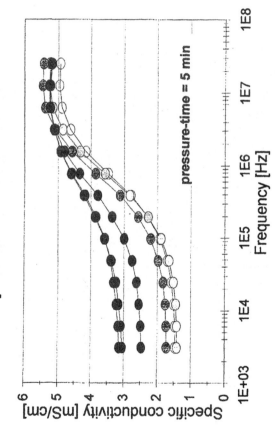

Figure 2.13 Frequency-dependent changes in the relative conductivity of cultured potato cells after pressure treatment.

thermal processes) can only be utilized at its fullest potential if the mechanisms involved are understood, essential kinetic data are accumulated, interactions between food constituents and pressure effects as well as shelf-life-dependent effects are studied exhaustively, and legislative issues are dealt with satisfactorily. There are tremendous challenges, but the benefits of fully utilizing the potential of high pressure rather than applying it to mimic existing processes seems to be worth it.

ACKNOWLEDGEMENTS

Parts of this work have been supported by the German Industrial Research Foundation (Grant No: AIF-FV 9918) and by the European Commission (Grant No AIR CTI-92-0296, FAIR-CT 96-1175).

REFERENCES

Angersbach, A., and Knorr, D. 1997. High electric field pulses as pretreatment to affect dehydration characteristics and rehydration properties of potato cubes. *Nahrung/Food* 41:194–200.

Angersbach, A., Heinz, V., and Knorr, D. 1997. Frequency dependent conductivity as a measure of the degree of permeabilization of cellular materials. *Lebensmittel- und Verpackungstechnik* (LVT) 42:195–200.

Anon. 1993. Use of high hydrostatic pressure in food processing. Overview. Outstanding Symposia in Food Science and Technology. *Food Technology.* 47 (6):149–172.

Balny, C., Hayashi, R., Heremans, K., and Masson, P. (Eds.) 1992. *High Pressure and Biotechnology.* Coloques INSERM/John Libbey Eurotext Ltd., Montrouge.

Barbosa-Cánovas, G. V., and Welti-Chanes, J. 1995. *Food Preservation by Moisture Control.* Lancaster, PA: Technomic Publishing Co. Inc.

Cheftel, J. C. 1995. High-pressure, microbial inactivation and food preservation. *Food Sci. Technol. Int.* 1:75–90.

Chlopin, G. W., and Tamman, G. 1903. Über den Einfluβ hober Drücke auf Mikroorganismen. *Z. Hyg. Infektionskrankheiten.* 45:171–204.

Dörnenburg, H., and Knorr, D. 1993. Cellular permeabilization of cultured plant tissues by high electric field pulses or ultra high pressure for the recovery of secondary metabolites. *Food Biotechnol.* 7:35–48.

Dörnenburg, H., and Knorr, D. 1997. Evaluation of elicitor and high pressure induced enzymatic browning utilizing potato (*Solanum tuberosum*) suspension cultures as model system for plant tissues. *J. Agric. Chem.* 45:4173-4177.

Eshtiaghi, M. N., and Knorr, D. 1993. Potato cubes' response to water blanching and high hydrostatic pressure. *J. Food Sci.* 58:1371–1374.

Eshtiaghi, M. N., and Knorr, D. 1996. High hydrostatic pressure thawing for the processing of fruit preparations from frozen strawberries. *Food Biotechnol.* 10:143–148.

Gould, G. W. 1995. *New Methods of Food Preservation.* Blackie Scientific, London.

Hayashi, R. 1993. *Use of High Pressure in Food.* San-Ei Publishing Co., Kyoto, Japan.

Hayashi, R. 1994. *Pressure Processed Food Research and Development.* San-Ei Publishing Co., Kyoto, Japan.

Hayashi, R., and Balny, C. 1997. *High Pressure Bioscience and Biotechnology.* Elsevier, Amsterdam.

Heinz, V. 1997. Wirkung hoher hydrostatischer Drücke auf das Absterbe- und Keimungsverhalten sporenbildender Bakterien am Beispiel *Bacillus subtilis* ATCC 9372. Ph.D. Thesis, Berlin University of Technology, Berlin.

Heinz, V., and Knorr, D. 1996. High pressure inactivation kinetics of *Bacillus subtilis* cells by a three-state-model considering distributed resistance mechanisms. *Food Biotechnol.* 10:149–161.

Hill, S. 1997. Squeezing the death out of food. *New Scientist.* 154 (2077):28–32.

Kalichevsky, M., Knorr, D., and Lillford, P. J. 1995. Potential food application of high-pressure effects on ice-water transitions. *Trends Food Sci. Technol.* 6:253–259.

Knorr, D. 1994. Plant cell and tissue cultures as model systems for monitoring the impact of unit operations on plant foods. *Trends Food Sci. Technol.* 5:328–331.

Knorr, D. 1995. Hydrostatic pressure treatment of food: Microbiology. In: Gould, G. W. (Ed.), *New Methods of Food Preservation.* Blackie Scientific, London, pp. 159–175.

Koch, H., Seyderhelm, I., Wille, P., Kalichevsky, M. T., and Knorr, D. 1996. Pressure-shift freezing and its influence on texture, colour, microstructure and rehydration behavior of potato cubes. *Nahrung/Food.* 40:125–131.

Labuza, T. P. 1994. Shifting food research paradigms for the 21st century. *Food Technol.* 48(12):50–56.

Messens, W., Van Camp, J., and Huyghebaert, A. 1997. The use of high pressure to modify the functionality of food proteins. *Trends Food Sci. Technol.* 8:107–112.

Mozhaev, V. V., Lange, R., Kudryashova, E. V., and Balny, C. 1996. Application of high hydrostatic pressure for increasing activity and stability of enzymes. *Biotechnol. Bioeng.* 52:320–331.

Ohshima, T., Ushio, H., and Koizumi, C. 1993. High pressure processing of fish and fish products. *Trends Foods Sci. Technol.* 4:370–375.

Sale, A. J. H., Gould, G. W., and Hamilton, W. A. 1970. Inactivation of bacterial spores by hydrostatic pressure. *J. Gen. Microbiol.* 60:323–334.

Smelt, J. P. M., Rigue, A. G. F., and Hayhurst, A. 1994. Possible mechanism of high pressure inactivation of microorganisms. *High Press. Res.* 12:199–203.

Stute, J. P. M., Estiaghi, M. N., Boguslawski, S., and Knorr, D. 1996. High pressure treatment of vegetables. In: Rohr, R., and Trepp, C. (Eds.), *High Pressure Chemical Engineering.* Elsevier, Amsterdam, S. 271–276.

Tauscher, B. 1995. Pasteurization of food by hydrostatic pressure: Chemical aspects. *Z. Lebensm. Unters. Forsch.* 200:3–13.

Weemaes, C., De Cordt, S., Gossens, M., Ludikhuyze, L., Hendrickx, M., Heremans, K., and Toback, P. 1996. High pressure, thermal and combined pressure-temperature stability of α-amylase from *Bacillus* species. *Biotechnol. Bioeng.* 50:49–56.

Wille, P., and Knorr, D. 1999. Use of *Solanum tuberosum* cell cultures for the evaluation of heat, pressure or high electric field pulse induced enzymatic browning in potato cells (in press).

CHAPTER 3

Modeling Water Absorption and Cooking Times of Black Beans (*Phaseolus vulgaris*) Treated with High Hydrostatic Pressure

ELBA SANGRONIS
ALBERT IBARZ
GUSTAVO V. BARBOSA-CÁNOVAS
BARRY G. SWANSON

INTRODUCTION

DOMESTIC consumption of beans is limited by the prolonged preparation times needed to soften beans to acceptable texture (Swanson, 1988). Because water absorption during soaking softens beans and reduces cooking times, beans are generally soaked for 10 to 16 h at room temperature prior to cooking. Development of flat sour and other bacterial growth is a problem associated with prolonged soaking times (Ogwal and Davis, 1994). Soaking beans at temperatures higher than 30°C (Abu-Ghannam and McKenna, 1997) or adding selected salts or chelating agents to soaking solutions (Del Valle et al., 1992) reduces soaking times and softens black beans. Alternative treatments such as vacuum (Sastry et al., 1985), gamma irradiation (Rao and Vakil, 1983), microwave (Abdul-Kadir et al., 1990), or ultrasonics (Uebersax et al., 1991) are effective in increasing water absorption and reducing cooking times of beans.

The Mattson cooker objectively determines cooking times (*CT*) or cookability of beans (Mattson, 1946). A modified version of the Mattson cooker using 25 beans instead of 100 beans is used extensively (Jackson and Varriano-Marston, 1981; De León et al., 1992; Hsieh et al., 1992; Nielsen et al., 1993). Chinnan (1985) automated the modified Mattson cooker. The number of beans with acceptable texture vs. time is plotted on log paper. Bean cookability or cooking time when 50% of beans attain an acceptable texture ($CT_{50\%}$) is calculated from the curve on log paper. Because the *CT* is a time-dependent variable, a kinetic model not previously studied may be useful to estimate the

31

effect of soaking and high hydrostatic pressure (HHP) treatment on cooking times of soaked or HHP-treated black beans.

Carbohydrate and protein bean constituents are responsible for water absorption and softening during soaking and cooking procedures (Hohlberg and Stanley, 1987). High hydrostatic pressure treatments modify the functionality of proteins and complex carbohydrates (Heremans, 1995). Therefore, the objectives of this study were (1) to determine the effect of HHP treatments on water absorption and cooking times of black beans; (2) to propose an empirical kinetic model predicting the water absorption of untreated and HHP-treated black beans; and (3) to propose an empirical kinetic model predicting the cooking times of soaked and HHP-treated black beans from results obtained with the modified Mattson bean cooker.

MATERIALS AND METHODS

Common beans (*Phaseolus vulgaris* L.) cv. Black Turtle Soup were acquired in Homedale, ID.

HIGH HYDROSTATIC PRESSURE TREATMENTS

A warm isostatic pressure system (Engineered Pressure Systems, Inc., Andover, MA) equipped with a cylindrical pressure chamber (height = 0.25 m, diameter = 0.1 m) was used for HHP treatments. The pressure medium was a 5% (v/v) Mobil Hydrasol 78 aqueous solution. Twenty grams of dry black beans were placed in polyethylene bags (7 cm × 5 cm) containing water (1:3 w/w). The bags were heat sealed. Three bags were placed in a larger polyethylene bag (23 cm × 25 cm) containing water. The larger bags were heat sealed and treated at 275, 410, 550, or 690 MPa, 25°C for 5 min. The come up times for 275, 410, 550, or 690 MPa were 2.7, 3.3, 4.5, and 5.5 min, respectively. Each treatment was duplicated, and assays were conducted in triplicate.

WATER ABSORPTION

Ten grams of black beans were soaked in 50 ml of water at 25°C for 24 h. Every hour, the soaked beans were blotted with absorbent paper to remove excess water, were weighed, and were placed back into the soaking water. Water absorption (m) was expressed on a dry weight basis and was calculated by the following equations:

$$m = \frac{\text{weight soaked beans} - \text{weight dry beans}}{\text{weight dry beans}}$$

therefore,

$$m = \frac{\text{g water}}{\text{g dry solid}}$$

Modification of the rate of water absorption by the black beans as an effect of HHP treatments was determined. High hydrostatic pressure-treated black beans were drained immediately after pressure treatments, were weighed, and were soaked. Water absorption during soaking was determined after 3, 6, 9, and 12 h. Correction for loss of solids during soaking or HHP treatments was not considered.

COOKING TIMES

Cooking times (*CT*) were determined using a modified Mattson cooker (Jackson and Varriano-Marston, 1981; Hsieh et al., 1992; Nielsen et al., 1993). The Mattson cooker is cylindrical and holds 25 beans for each experiment. Each plunger weighing 90 g ends in a stainless-steel rod (0.16 cm diameter) that rests on the beans. As cooking progresses and the beans soften, the plunger penetrates the seed coat and cotyledons approximately 2 to 3 cm. When each bean softens to an acceptable texture, the plunger penetrates the bean, and the time is recorded. The curve of beans with acceptable softened textures as a function of cooking time is sigmoidal. The *CT* of beans is the time in minutes required for penetration of 50% of the beans (Jackson and Varriano-Marston, 1981). Cooking times of HHP-treated black beans were determined immediately after pressure treatment, while the *CT* of untreated raw beans was determined after soaking for 3, 6, 9, or 12 h.

STATISTICAL ANALYSIS

Statistical tests were performed using Minitab® Statistical software, version Release 12 (Minitab, Inc., 1997). Nonlinear regression was used to fit the proposed model of water absorption and cooking times. Significance was defined as $P \leq 0.05$.

RESULTS AND DISCUSSION

WATER ABSORPTION

The water absorption curve of soaked raw beans (Figure 3.1) exhibited the characteristic pattern of black beans: initial rapid water uptake followed by saturation after 10 to 16 h of soaking (Sefa-Dedeh et al., 1978; Giami and

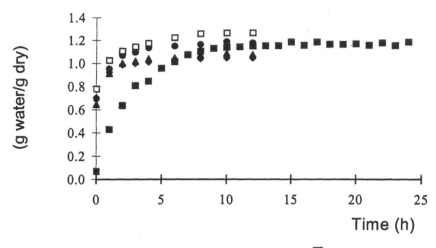

Figure 3.1 Water imbibition of untreated raw beans soaked for 24 h ■ and black beans treated at 275 MPa ●, 410 MPa ▲, 550 MPa or 690 MPa ◆.

Okwechime, 1993; Berrios-Silva, 1995). The moisture content of dry black beans was 0.069 g/g (dwb). Moisture content of dry beans after HHP treatment for 5 min was from 10- to 12-fold higher than the initial moisture content of black beans. At the saturation of the water absorption curve (Figure 3.1), beans treated at 550 MPa contained significantly more water than beans treated at 275, 410, or 690 MPa was observed. The saturation of HHP-treated beans was attained after 5 to 6 h of soaking, whereas unsoaked raw black beans required 10 to 12 h to reach saturation. High hydrostatic pressure treatments increased the rate of water absorption and the amount of water imbided. The time necessary to reach the saturation did not depend on level of pressure applied.

KINETIC MODEL FOR WATER ABSORPTION

Water absorption can be described in two phases: a first phase (a) when the water is absorbed and a consecutive phase (b) when the water is released. Because the amount of water during soaking is always greater than the solid content of black beans, a kinetic of order zero was proposed for the first phase, whereas, a kinetic of first order was proposed to describe the desorption phase. Therefore,

$$\text{(a) } A + H_2O \xrightarrow{k_0} A - H_2O \qquad (n = 0)$$

$$\text{(b) } A - H_2O \xrightarrow{k_1} A + H_2O \qquad (n = 1)$$

k_0 and k_1 were the kinetic constant of the first and second phase, respectively. The variable m was already defined as

$$m = \frac{g\ water}{g\ dry\ solid} \tag{1}$$

Therefore, the variation of m with the time can be expressed by the derivative

$$\frac{dm}{dt} = k_0 - k_1\ m \tag{2}$$

when $t = 0$, $m = m_0$. Then, Equation (2) can be expressed as

$$m = \frac{k_0}{k_1} - \left(\frac{k_0}{k_1} - m_0\right) \exp(- k_1 t) \tag{3}$$

Then, the ratio k_0/k_1 was defined as K

$$m = K - (K - m_0) \exp(-k_1 t) \tag{4}$$

Equation (4) varies with the soaking times, but at the equilibrium level m is a constant

$$m_{Eq} = K \tag{5}$$

$$k_0 = k_1\ m_{Eq} \tag{6}$$

A non-linear regression was used to fit the kinetic model. The predictive capability of Equation (4) was demonstrated (Steele et al., 1997). Differences between experimental and predictive values were not significant at $P \leq 0.05$. Regression coefficients (R^2) ranged from 0.98 to 0.99 P (Table 3.1). The ratio

TABLE 3.1. Kinetic Constants Calculated from Equation (4),* and the Regression Coefficient of the Proposed Model for Untreated and HHP* Treated Black Beans.

Treatment	$m_{Eq} = k_0/k_1$ (g water/g dry)	k_1	k_0	R^2
Soaked	1.171	0.344	0.403	0.9959
690 MPa	1.031	0.986	1.017	0.9845
550 MPa	1.256	0.563	0.707	0.9809
410 MPa	1.065	0.925	0.085	0.9881
275 MPa	1.170	0.721	0.844	0.9921

*$m = K - (K - m_0) \exp(- k_1 t)$.
R_2: regression coefficient ($P \leq 0.05$).
K_0: kinetic constant of water absorption. K_1: kinetic constant of water desorption.

TABLE 3.2. Experimental Cooking Times of Unsoaked, Soaked, or HHP-Treated Black Beans.

Treatment	Cooking Times (min) Mean ± SD
Unsoaked	44.1 ± 1.2[a]
Soaked 3 h	30.9 ± 3.8[b]
Soaked 6 h	21.6 ± 0.1[c]
Soaked 9 h	21.8 ± 1.1[c]
Soaked 12 h	19.6 ± 1.4[c]
275 MPa	31.9 ± 3.7[b]
410 MPa	30.8 ± 2.5[b]
550 MPa	28.3 ± 1.3[b]
690 MPa	27.2 ± 0.9[b]

Means of triplicate. Means with different letters are significantly different ($P \leq 0.05$).

k_0/k_1 was always greater than 1 and indicated the amount of water absorbed at the equilibrium condition. The m_{Eq} of untreated black beans and black beans treated at 275 MPa were the same, while the black beans treated at 550 MPa exhibited the greatest m_{Eq}. The m_{Eq} may indicate the ability of black beans to absorb water.

COOKING TIMES

The mean cooking time (CT) of unsoaked black beans was 44 min. Soaking for 3 to 12 h reduced the CT of unsoaked black beans to 40 and 27 min, respectively, representing a 30% to 56% reduction, respectively (Table 3.2). Soaking and cooking times were inversely related. Hence, beans soaked for 3 h required significantly more cooking to attain acceptable texture than black beans soaked for 6, 9, or 12 h. The CT of beans soaked for 6, 9, or 12 h did not differ significantly. The CT of black beans treated at 690 MPa was 27 min, a 39% reduction of CT in comparison to untreated raw black beans (44 min). Pressure treatment of 550 MPa reduced the CT of untreated black beans by 36%. When HHP treatments were 410 or 275 MPa, the reduction in CT was 30% and 25%, respectively. Differences among the CT of beans treated at 275, 410, 550, or 690 MPa were not significant. Treating black beans at 275, 410, 550, or 690 MPa for 5 min exhibited CT equivalent to soaking beans for 3 h.

Moisture content of the beans before cooking and CT were inversely and significant correlated ($r = -0.97$), meaning that black beans with the greatest initial moisture content exhibited the shortest CT. Sefa-Dedeh and Stanley (1979) correlated the hardness of soaked beans and the amount of water absorbed. The correlation coefficients ranged from -0.85 to -0.99. Legumes with greater water absorption during soaking require less CT than legumes

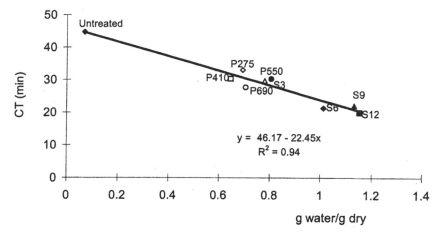

Figure 3.2 Cooking times (*CT*) as a function of initial water content of black beans after soaking or HHP treatment (S = soaking for the given time; P = HHP treated at the given pressure).

with little water absorption (Jackson and Varriano-Marston, 1981; Abdul-Kadir et al., 1990). Determination of linear regression between initial moisture content and *CT* (Figure 3.2) indicates the initial moisture content of soaked or HHP-treated black beans before cooking accurately predicts ($R^2 = 0.94$) the *CT* using the following equation:

$$CT = 46.17 - 22.45\ X$$

where *X* is the moisture content of beans before cooking.

The relationship established by linear regression between initial moisture content suggests a practical application. Knowing the initial moisture content of black beans after soaking or HHP treatment, the *CT* can be calculated from the equation obtained with linear regression.

High hydrostatic pressure treatment of 275, 410, 550, and 690 MPa or 25°C for 5 min increases the rate of water absorption and amount imbibed. Saturation was attained in 5 to 6 h, whereas black beans soaked in water at 25°C required 10 to 12 h to attain saturation. Cooking time reduction of black beans treated at any of the four pressures for 5 min was equivalent to the *CT* of black beans soaked for 3 h. The results suggest that HHP treatment is an alternative process to improve the efficiency of bean preparation. High hydrostatic pressure is advantageous in respect to other alternatives previously reported. Soaking beans in water for several hours consumes energy and produces a large amount of effluents. High hydrostatic pressure is a low energy process without environmental pollution and will eliminate the use of chemical additives to reduce the prolonged cooking times of black beans.

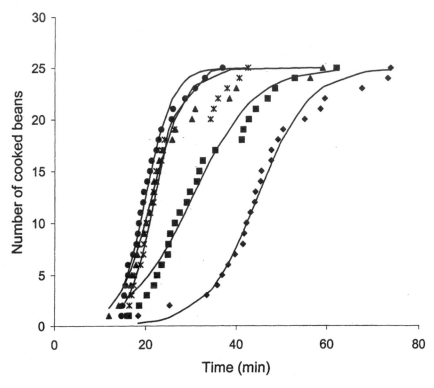

Figure 3.3 Experimental (points) and predicted (lines) cooking times (*CT*) of unsoaked raw beans ◆ and black beans soaked for 3 h ■, 6 h ▲, 9 h * or 12 h ●.

MODELING COOKING TIMES

Graphs of numbers of beans with acceptable texture vs. time give an S-shaped (sigmoid) curve (Figures 3.3 and 3.4). The number of beans penetrated by the plunger accumulates with time. Thus, the sigmoid curve obtained suggests a cumulative function that can be adjusted with a kinetic model (Froment and Bischoff, 1990). The following expression describes bean cooking:

$$A \xrightarrow{k} B \tag{7}$$

where A is the number of beans that are not soft enough to be penetrated by the plunger, B is the number of beans with acceptable softness, and k is the kinetic constant (Froment and Bischoff, 1990). The accumulation of beans with the acceptable texture over time is expressed as

$$\frac{dB}{dt} = kAB \qquad (8)$$

The initial number of beans is A_0, then, at any time

$$A_0 = A + B \qquad (9)$$

Therefore, Equation (8) is expressed as

$$\frac{dB}{dt} = k(A_0 - B)B \qquad (10)$$

Integrating Equation (10) at the condition limit, when $t = 0$ and $B = 0$, results in the following equation:

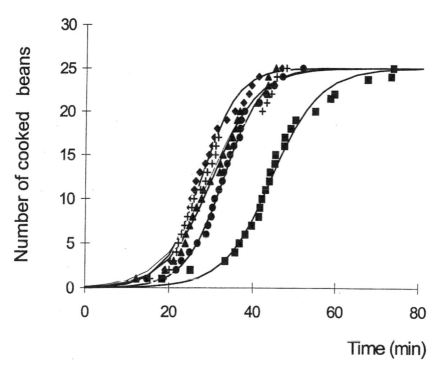

Figure 3.4 Experimental (points) and predicted (lines) cooking times (*CT*) of unsoaked-untreated black beans ■ and beans treated at 275 MPa ●, 440 MPa ▲, 550 MPa + or 690 MPa ◆.

$$B = \frac{A_0}{1 + A_0 \exp[- A_0 \, kt)]} \qquad (11)$$

The accumulation of beans with acceptable softness as a time function is expressed in Equation (11).

The cooking time curves (Figures 3.3 and 3.4) indicated that the first bean was penetrated at 10 to 12 min. The time required to soften and penetrate the first bean by the plunger is the induction period (τ). Then, Equation (11) can be described as a function of τ:

$$B = \frac{A_0}{1 + A_0 \exp[- A_0 \, k(t - \tau)]} \qquad (12)$$

Cooking time or CT_{50} is defined as the time when 50% of beans have acceptable softness (Mattson, 1946; Jackson and Varriano-Marston, 1981). Then, $CT_{50} = A_0/2$, and $t = CT$. Then:

$$\frac{A_0}{1 + A_0 \exp[- A_0 \, k(CT - \tau)]} = \frac{A_0}{2} \qquad (13)$$

Mathematically:

$$\frac{1}{2} + \frac{A_0}{2} \exp[-A_0 \, k(CT - \tau)] = 1 \qquad (14)$$

and:

$$\exp[-A_0 \, k(CT - \tau)] = \frac{1}{A_0} \qquad (15)$$

therefore;

$$(CT - \tau) = - \frac{1}{A_0 k} \ln \frac{1}{A_0} \qquad (16)$$

So that:

$$CT = - \frac{1}{A_0 k} \ln \frac{1}{A_0} + \tau \qquad (17)$$

$$CT = \frac{\ln A_0}{A_0 k} + \tau \qquad (18)$$

TABLE 3.3. Parameters of Cooking Times Calculated Using Equation (13)* for Unsoaked, Soaked, and HHP-Treated Black Beans and the Regression Coefficient.

Treatment Type	$k10^{-3}$	τ (min)	CT (min)	R^2
Unsoaked	6.9 ± 1.2	26.0 ± 3.0	44.67	0.9835
Soaked 3 h	5.8 ± 1.1	8.0 ± 4.0	30.21	0.9694
Soaked 6 h	11.0 ± 2.0	9.6 ± 2.4	21.31	0.9735
Soaked 9 h	14.0 ± 4.0	12.8 ± 2.5	22.00	0.9530
Soaked 12 h	13.8 ± 2.3	10.6 ± 1.5	19.93	0.9797
275 MPa	8.9 ± 0.8	18.5 ± 1.3	32.97	0.9944
410 MPa	7.4 ± 0.7	12.9 ± 1.8	30.31	0.9908
550 MPa	6.9 ± 1.2	11.0 ± 3.0	29.67	0.9748
690 MPa	9.4 ± 1.3	13.8 ± 1.9	27.65	0.9843

*$CT = \tau + 0.1288/k$.
k.: kinetic constant, τ: induction time. CT: cooking times. R^2: regression coefficient.

Because A_0 is equal to 25 beans, and ln 25/25 = 0.1288, then the CT can be expressed as

$$CT = \tau + \frac{0.1288}{k} \tag{19}$$

The predictive capability of Equation (19) was demonstrated (Figures 3.3 and 3.4) using non-linear regression (Steele et al., 1997). Differences between experimental and predictive values were not significant. Regression coefficients (R^2) ranged from 0.95 to 0.99 (Table 3.3). The τ, k, and CT values varied with soaking times or HHP treatment applied to beans prior to cooking.

CONCLUSIONS

High hydrostatic pressure treatments increased the rate of water absorption as well the amount of water imbibed by black beans compared to untreated black beans. Saturation in HHP-treated black beans was attained much faster than in untreated black beans. High hydrostatic pressure treatment reduced the cooking times of black beans. The empirical kinetic models significantly predicted the water absorption and cooking times of untreated and HHP-treated black beans. Results indicate that HHP may be useful technology to reduce prolonged soaking and cooking time of black beans.

REFERENCES

Abdul-Kadir, R., Bargman, T. J., and Rupnow, J. 1990. Effect of infrared heat processing on rehydration rate and cooking of *Phaseolus vulgaris* (Var. Pinto). *J. Food Sci.* 55:1472–1473.

Abu-Ghannam, N., and Mckenna, B. 1997. Hydration kinetics of red kidney beans (*Phaseolus vulgaris* L.). *J. Food Sci.* 62:520–523.

Berrios-Silva, J. 1995. Storage of black beans (*Phaseolus vulgaris.* L): Composition structure, and bromelin hydrolysis of globulin G1. Ph.D. Dissertation. Washington State University, Pullman. p. 72.

Chinnan, M. S. 1985. Development of a device for quantifying hard to cook phenomenon in cereal legumes. *Trans. ASAE* 28 (1):335–339.

De Leon, L. F., Elias, L. G., and Bressani, R. 1992. Effect of salt solution on the cooking time, nutritional and sensory characteristics of common bean (*Phaseolus vulgaris*). *Food Res. Int.* 25:131–137.

Del Valle, J. M., Stanley, D. W., and Bourne, M. C. 1992. Water absorption and swelling in dry beans seed. *J. Food Proc. Pres.* 16:75–98.

Froment, G. F., and Bischoff, K. B. 1990. Chemical reactors analysis and design. New York: John Wiley & Sons, Inc. p. 61.

Giami, S. Y., and Okwechime, U. I. 1993. Physicochemical properties and cooking quality of four new cultivars of Nigerian cowpeas (*Vigna inguiculate* L Walp). *J. Sci. Food Agric.* 63:281–286.

Heremans, K. 1995. High pressure effect on biomolecules. In *High Pressure Processing of Food.* Ledward, D. A., Johnston, D. E., Earnshaw, R. E., and Hasting, A. P. M., eds. Nottingham, UK: Nottingham University Press. p. 81.

Hohlberg, A. I., and Stanley, D. W. 1987. Hard to cook defect in black beans. Protein and starch considerations. *J. Agric. Food. Chem.* 35:571–576.

Hsieh, H. M., Pomeranz, Y., and Swanson, B. G. 1992. Composition, cooking time and maturation of azuki (*Vigna angularis*) and common beans (*Phaseolus vulgaris*). *Cereal Chem.* 69:244–248.

Jackson, G. M., and Varriano-Marston, E. 1981. Hard-to-cook phenomenon in beans: Effect of accelerated storage on water absorption and cooking time. *J. Food Sci.* 46:799–803.

Mattson, S. 1946. The cookability of yellow peas. A colloid-chemical and biochemical study. *Acta. Agr. Suecana* 2:185–188.

Minitab, Inc. 1997. Minitab Statistical Softwere. Release 12. State College, PA.

Nielsen, S., Brandt, W. E., and Singh, B. B. 1993. Genetic variability for nutritional composition and cooking time of improved cowpea lines. *Crop Sci.* 33:469–472.

Ogwal, M. O., and Davis, D. R. 1994. Rapid rehydration methods for dried beans. *J. Food Sci.* 59:611–612, 654.

Rao, V. S., and Vakil, U. K. 1983. Effects of gamma-irradiation on flatulence causing oligosaccharides in green gram (*Phaseolus areus*). *J. Food Sci.* 48:1791–1793.

Sastry, S. K., McCafferty, F. D., Murakami, E. G., and Kuhn, G. D. 1985. Effects of vacuum hydration on the incidence of splits in canned kidney beans. *J. Food Sci.* 50:1501–1503.

Sefa-Dedeh, S., and Stanley D. W. 1979. Textural implications of the microstructure of legumes. *Food Technol.* 33:77–83.

Sefa-Dedeh, S., Stanley, D. W., and Voisey P. W. 1978. Effect of the storage time and conditions on the hard-to-cook defect in cowpea (*Vigna unguiculata*). *J. Food Sci.* 44:790–795.

Steele, R. G. D., Torrie, J. H., and Dickey, D. A. 1997. *Principles and Procedures of Statistic. A Biometrical Approach.* 3rd edition. pp. 138–177. New York: McGraw-Hill Book Co.,

Swanson, B. G. 1988. Beans are good food. *Michigan Bean Digest* 12:14–15.

Uebersax, M. A., Ruengsakulrach, S., and Occena, L. G. 1991. Strategies for processing dry beans. *Food Technol.* 45:104–108, 110–111.

CHAPTER 4

High Hydrostatic Pressure Inactivation of Selected Pectolytic Enzymes

ALBERT IBARZ
ELBA SANGRONIS
CARMEN GONZÁLEZ
GUSTAVO V. BARBOSA-CÁNOVAS
BARRY G. SWANSON

INTRODUCTION

Textural changes in fruits and vegetables are related to biochemical and biophysical alterations that occur during ripening and processing. The firmness and softening of tissues is often correlated with the activity of pectolytic enzymes (Verlinden and De Baerdemaeker, 1997). Thermal processes, such as blanching, cooking, and canning, applied to fruits and vegetables are designed to inactivate enzyme and maximize the retention of textural properties (Verlinden et al., 1996).

In fruit products such as oranges, guavas, and tomatoes, the texture and consistency are the most important quality characteristics determining preference by consumers (Lin and Yen, 1995). The viscosity of juices depends on suspended solids and the total amount of pectic substances. Endogenous pectolytic enzymes degrade pectic substances and diminish juice viscosity, causing the quality of the products to decrease (López et al., 1997).

Enzyme inactivation is traditionally carried out by thermal processes, but high temperature can have a negative impact on the sensory characteristics of fruits and vegetables. HHP is emerging as a revolutionary nonthermal process successfully used in food processing. High pressure processing produces minimal changes in the flavor, color, or taste of food. Moreover, pressurization modifies the tertiary and quaternary structures of proteins, which can explain the inactivation of microorganisms and extended shelf life of food preserved with the high-pressure process (Cheftel, 1995; Ashie and Simpson, 1996). Pressurization also modifies the conformation of the active site and activities of enzymes. Even though some enzymes are inactivated by high-pressure treatments, selected enzymes are also activated (Lin and Yen, 1995; Seyderhelm et al., 1996).

In freshly squeezed and non-pasteurized Satsuma mandarin juices, Ogawa et al. (1990) observed that for pressurization greater than 300 MPa pectinesterase activity decreased when pressure and the soluble solids content increased. Levels of pressure greater than 400 MPa for 10 min at 30°C reduced the pectinesterase activity in guava juice with 3° Brix and pH 4.7. At 600 MPa, the pectinesterase inactivation was 33% (Lin and Yen, 1995), while at pH 3.9, the inactivation was 60%. For guava juices with 12° Brix at pH 3.9 processed at 25°C, the pectinesterase activity was reduced 40%. So, when guava juices were treated under HHP, the pectinesterase inactivation increased with an increase of acidity and soluble solids content. According to Seyderhelm et al. (1996), pressures greater than 800 MPa at 45°C inactivated pectinesterase. Furthermore, an increase in soluble solids exercised a baroprotective effect on pectinesterase inactivation. Cano et al. (1997) observed a loss of pectin methylesterase activity (25%) in freshly squeezed orange juices treated with 200 MPa at 30°C.

The effectiveness of high-pressure processing on enzyme activity depends on factors such as pH, temperature, enzyme concentration, and medium condition (Porretta et al., 1994; Lin and Yen, 1995; Cano et al., 1997). Because the effect of high pressure on polygalacturonase has not been studied, the objectives of this study were (1) to investigate pressure-induced inactivation of endo- and exo-polygalacturonase and pectinmethylesterase; (2) to identify the importance of pH and temperature on the inactivation degree of these pectolytic enzymes; and (3) to study the time effect at the highest inactivation pressure.

MATERIALS AND METHODS

Endo- and exo-polygalacturonases (PG) (EC 3.2.1.15) from *Aspergillus niger* and pectin methylesterase (PME) from orange peel were acquired from Sigma Chemical Co. (St. Louis, MO). Three separate enzyme solutions were prepared, with 0.30 units of exo-PG per ml, 0.25 units per ml of endo-PG, and 0.25 units per ml of PME. For each enzymatic solution, four pHs were assayed. The enzymatic solutions were adjusted at pH 3.3, 4.0, 5.5, and 7.0. The activity of each enzyme used was defined according to Sigma Chemical Co. One activity unit of exo-PG liberated 1.0 μmole of galacturonic acid from polygalacturonic acid per minute, while one activity unit of endo-PG released 1.0 μmole of reducing sugar assayed as D-galacturonic acid from polygalacturonic acid per minute. One activity unit of PME released 1.0 miliequivalent of D-galacturonic acid from pectin per minute.

HIGH HYDROSTATIC PRESSURE TREATMENTS

Five milliliters of each enzymatic solution were placed in plastic bags (2 cm × 3 cm) that were heat sealed free of air bubbles, treated with HHP, and stored in ice water until enzymatic activity determination. The pressures applied were 138, 275, 413, 550, and 690 MPa for 2, 5, 10, 20, and 30 minutes. For kinetic inactivation, the selected conditions were 690 MPa at 25°C and 50°C. All treatments were conducted in an EPSI high hydrostatic press (Engineered Pressure Systems, Inc., Andover, MA) with a cylindrical pressure chamber (height: 0.25 m, diameter: 0.10 m). A 5% Mobil Hydrasol 78 water solution was used as the pressure medium.

ENZYME ACTIVITY

Pectin methylesterase activity was determined by the spectrophotometric method (Vilariño et al., 1993). One volume of 0.017 g/100 ml bromocresol green solution in distilled water was added to 10 volumes of citrus pectin solution in 0.5% (w/v) distilled water. The pH of the solution was adjusted to 5.1 with 0.01 N NaOH solution. The enzyme reaction started when 0.5 μl pectin methylesterase solution was added to 1.95 ml of pectin-bromocresol green solution. The reaction was continuously monitored at 618 nm in a Hewlett Packard spectrophotometer (Palo Alto, CA) model 8452A. The PME activity was expressed as a decrease in absorbance per minute.

Endo- and exo-PG activity were determined by the Pressey and Avants method (1978). For assay endo-PG activity, 0.4 ml of 0.2 mM sodium acetate buffer at pH 4.0 and 1 ml of endo-PG were mixed. For exo-PG activity, 0.4 ml of 0.2 mM sodium acetate buffer at pH 5.5, 0.1 ml of 0.02 M $CaCl_2$ and 1 ml of exo-PG solution were mixed. The endo- and exo-PG activity started when 1 ml of 1% polygalacturonic acid was added. The pH of this solution was previously adjusted to 4.0 for endo-PG activity and 5.5 for exo-PG activity. After 24 hours, the reducing groups liberated were determined (Nelson, 1944). A standard curve was made using galacturonic acid. The enzymatic activity was expressed as a relative value referred to untreated enzyme.

STATISTICAL ANALYSIS

Analysis of variance (ANOVA), LSD, and modeling were conduced using STATGRAPHICS 7.0 (1993). The level of significance was 95%.

RESULTS AND DISCUSSION

INACTIVATION OF PECTOLYTIC ENZYMES AT DIFFERENT PH

For all pressures assayed at 25°C for 10 minutes, the highest exo-PG inactivation was obtained at pH 5.5, while the lowest inactivation was observed at pH 4.0 (Figure 4.1). At pH 3.3, 4.0, 5.5, and 7.0, a slight decrease in activity was observed when the pressure was increased. At pH 5.5 and 690 MPa, the activity of exo-PG reached the maximum inactivation, but, in general, at the selected conditions, the exo-PG activity was not highly affected. According to Hoover (1993), the polyphenol-oxidase from peaches was weakly affected by high-pressure treatment, as it showed similar behavior to the exo-PG.

The endo-PG was more sensitive to pressure treatment than exo-PG. The endo-PG activity decreased when pressure increased (Figure 4.2). The minimum activity of endo-PG was obtained when the pressure was increased to 550 or 690 Mpa, and the medium pH was 3.3. Only 12% of the activity was reached at the latter conditions. However, when the medium pH was 4.0, 5.5, and 7.0 at pressures of 550 or 690 MPa, the activity observed was 60% to 65%.

The PME showed little inactivation (Figure 4.3). However, when the pressure increased to 690 MPa at a pH of 3.3, the PME showed 16% inactivation.

INACTIVATION KINETICS OF PECTOLYTIC ENZYMES AT 690 MPA

According to previous results, the greatest inactivation of PME and endo- and exo-PG was achieved at 690 MPa. Therefore, the kinetic study was performed at this pressure. The effect of treatment time on pectolytic enzyme inactivation followed a first-order reaction expressed as the equation:

$$A = A_0 \exp(- kt) \tag{1}$$

where A and A_0 are the relative activity of enzyme at a given time and initial time, respectively, and k is a first kinetic constant for enzyme inactivation. The kinetic constant values for the different experiments are presented in Table 4.1. The k values obtained by fitting the experimental data to Equation (1) were significant at $P \leq 0.05$. A high kinetic constant value indicates a good inactivation. The k values for exo-PG at 50°C and all pH values were significantly greater than the k values obtained at 25°C. At pH 4.0 and 50°C, the k value was largest at 0.028 min^{-1}. The kinetic study demonstrated that the greatest inactivation for endo-PG was at 50°C and pH 7.0, and PME behaved similarly. Exceptionally, endo-PG at pH 3.3 presented a marked inactivation (Figure 4.4). At 25°C, only 2 min were needed to achieve 87% of endo-PG inactivation. At 50°C, the inactivation was similar to 25°C. How-

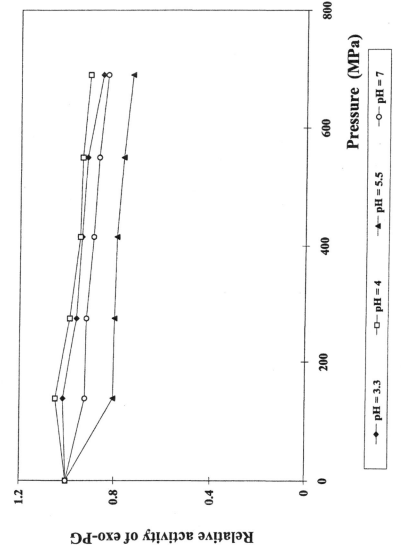

Figure 4.1 Effect of pH and pressure on the inactivation of exo-PG for 10 minutes at 25°C.

pH = 3.3   pH = 4   pH = 5.5   pH = 7

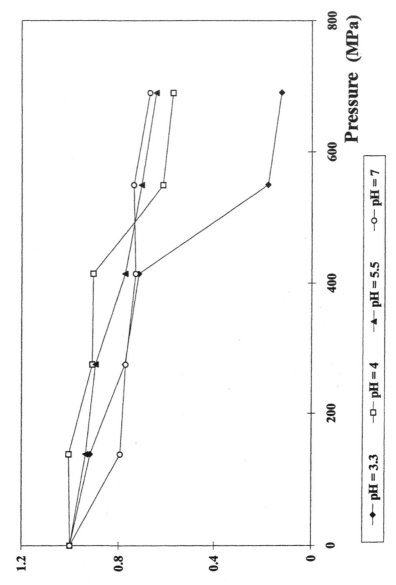

Figure 4.2 Effect of pH and pressure on the inactivation of endo-PG for 10 minutes at 25°C.

◆— pH = 3.3   □— pH = 4   ▲— pH = 5.5   ○— pH = 7

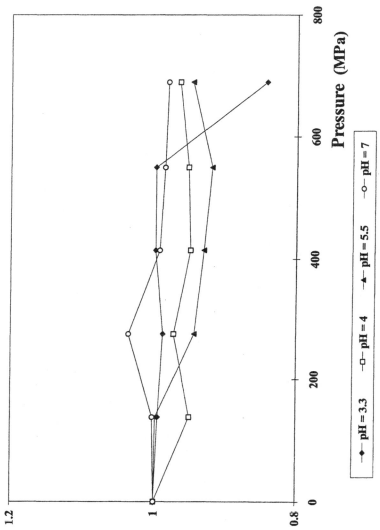

Figure 4.3 Effect of pH and pressure on the inactivation of PME for 10 minutes at 25°C.

49

TABLE 4.1. Kinetic Constant Values for the Inactivation of Pectolytic Enzymes at 690 MPa.

Enzyme	pH	k (min^{-1})	
		25°C	50°C
Exo-PG	3.3	0.007[a]	0.018[b]
	4	0.005[c]	0.028[d]
	5.5	0.010[e]	0.015[f]
	7	0.011[e]	0.014[f]
Endo-PG	3.3	—	—
	4	0.013[g]	0.017[h]
	5.5	0.021[m]	0.024[p]
	7	0.015[f]	0.052[k]
PME	3.3	0.003[l]	0.005[c]
	4	0.003[l]	l0.004[l]
	5.5	0.003[l]	0.003[l]
	7	0.006[c]	0.007[a]

Superscript means having different letters are significantly different ($P \leq 0.05$).

ever, after 2 min at 690 MPa for both temperatures, the inactivation of endo-PG was independent of time.

TEMPERATURE AND PRESSURE INACTIVATION OF PECTOLYTIC ENZYMES

The effect of pressure on exo-PG activity for three temperatures (25, 40, and 50°C) at pH 7.0 and for 10 min processing time is illustrated in (Figure 4.5). The activity decreased when pressure increased. Conditions such as 50°C and 690 MPa induced a maximum inactivation of 42%. The behavior of endo-PG is presented in Figure 4.6. The endo-PG demonstrated less barostability at pressures greater than 275 MPa combined with 40°C. At 275 MPa, the activity of endo-PG was 40%, while at 25°C and 275 MPa, the activity was 78% of the initial value. Pressures larger than 275 MPa did not increase the inactivation observed at 275 MPa. The behavior of PME is presented in Figure 4.7. At 138 MPa, a marked decrease of PME activity was observed at 40°C and 50°C. Pressures larger than 138 MPa did not increase the inactivation observed at 138 MPa.

The inactivation of endo-PG and PME at 40°C was greater than the inactivation at 50°C. According to Lin and Yen (1995), maximum enzyme activity is achieved at about 50°C, the temperature at which the activity of the two enzymes is maximum about 50°C, making activity reduction using HHP difficult. This fact explains why the activity reduction at 50°C is less than at 40°C (i.e., the pressure has to overcome the favoring action of the temperature to reduce the enzyme activity).

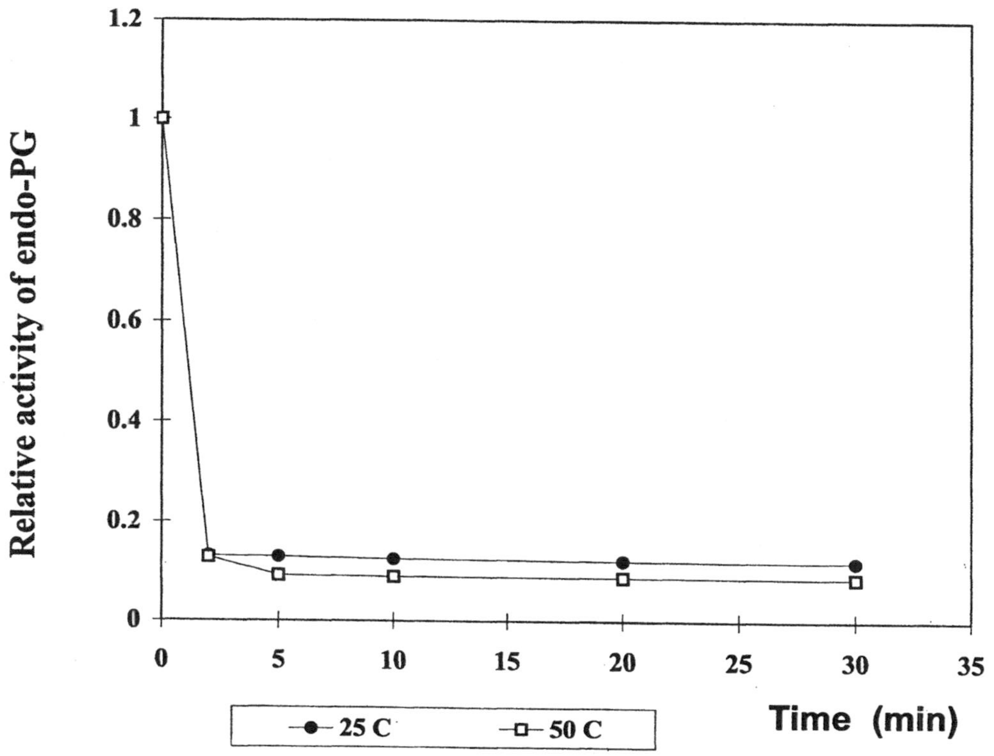

Figure 4.4 Effect of temperature and processing time on the inactivation of endo-PG at 690 MPa and pH 3.3.

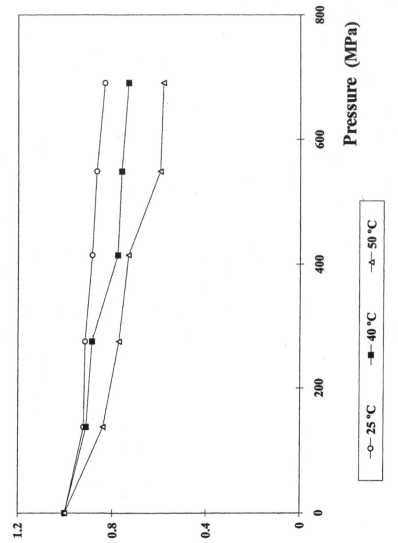

Figure 4.5 Effect of temperature and pressure on the inactivation of exo-PG for 10 minutes at pH 7.

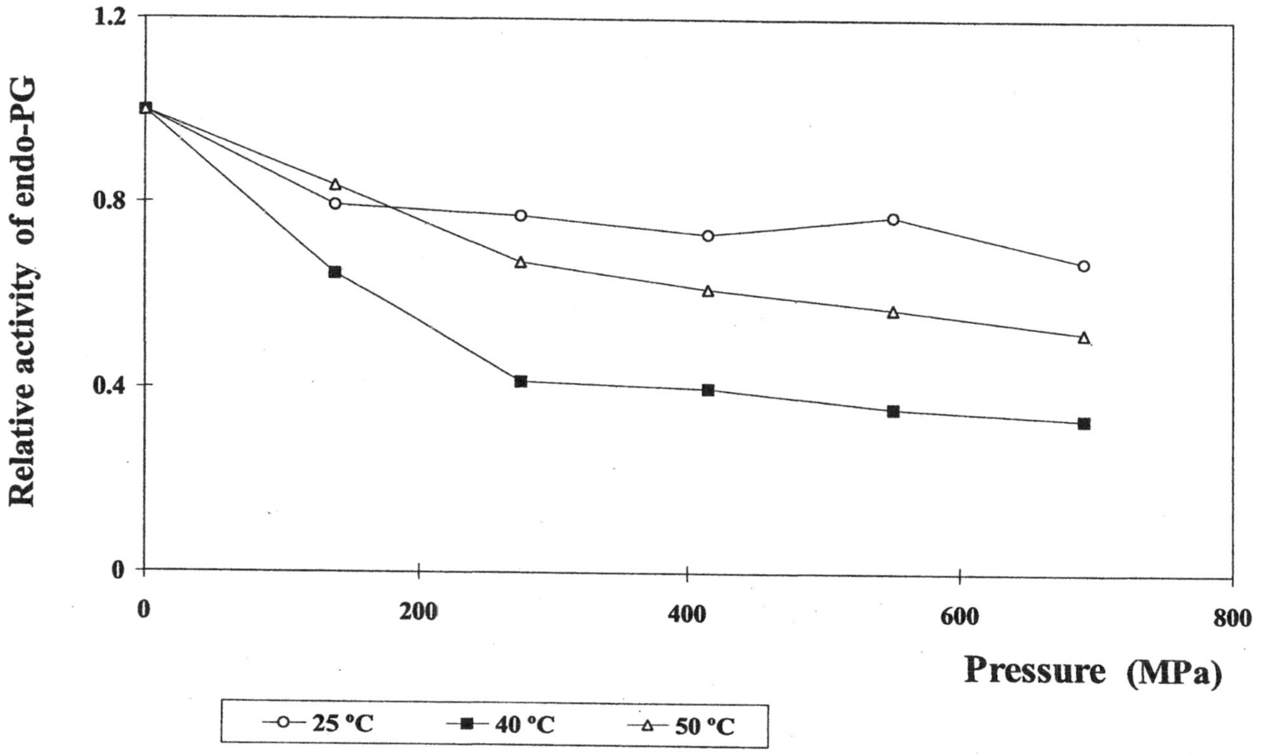

Figure 4.6 Effect of temperature and pressure on the inactivation of endo-PG for 10 minutes at pH 7.

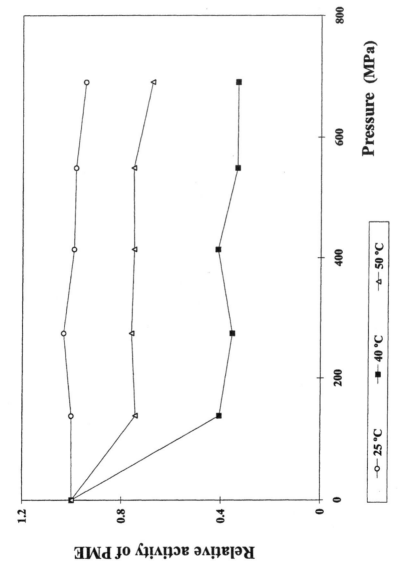

Figure 4.7 Effect of temperature and pressure on the inactivation of PME for 10 minutes at pH 7.

25 °C   40 °C   50 °C

54

CONCLUSION

Partial inactivation of PME and exo- and endo-PG was observed by HHP treatments, but inactivation also depended on medium pH, processing temperature, and processing time. Therefore, the combination of factors must be considered to inactivate pectolytic enzymes. The exo-PG and PME were found to be more barostable than endo-PG. The barostability of the three enzymes decreased when pressure was combined with temperatures of 40°C and 50°C. At 690 MPa and 25°C and 50°C, the variation of activity with time exhibited a first-order kinetic meaning that activity decreased exponentially when the time increased.

ACKNOWLEDGEMENT

The author A. Ibarz is very grateful to CICYT (Ministerio de Educación y Ciencia-Spain) (PR95-410) for the financial support that permitted his stay at Washington State University.

REFERENCES

Ashie, I. N. A., and Simpson, B. K. 1996. Application of high hydrostatic pressure to control enzyme related fresh sea food texture deterioration. *Can. Inst. Food Sci. Technol. J.* 19:569–575.

Cano, M. P., Hernández, A., and De Ancos, B. 1997. High pressure and temperature effects on enzyme inactivation in strawberry and orange products. *J. Food Sci.* 62:85–88.

Cheftel, J. C. 1995. High-pressure, microbial inactivation and food preservation. *Food Sci. Technol. Int.* 1:75–80.

Hoover, D. G. 1993. Pressure effects on biological systems. *Food Technol.* 47 (6):150–155.

Lin, H. T., and Yen, G. C. 1995. Effect of high hydrostatic pressure on the inactivation of enzymes and sterilization of guava juice. *J. Chinese Agric. Chem. Soc.* 33:18–29.

López, P., Sánchez, A. C., and Vercet, A. 1997. Thermal resistance of tomato polygalacturonase and pectinmethylesterase at physiological pH. *Z. Lebens. U. Lotssh.* 204:146–150.

Nelson, N. 1944. A photometric adaptation of the Somogyi method for the determination of glucose. *J. Biol. Chem.* 153:375–380.

Ogawa, H., Fukuhisa, K., Kubo, Y., and Fukumoto, H. 1990. Pressure inactivation of yeasts, molds, and pectinesterase in Satsuma mandarin juice: Effects of juice concentration, pH, and organic acids, and comparison with heat sanitation. *Agric. Biol. Chem.* 54:1219–1225.

Porretta, S., Birzi, A., Ghizzoni, C., and Vicini, E. 1994. Effects of ultra-high hydrostatic pressure treatments on the quality of tomato juice. *Food Chem.* 52(1):35–41.

Pressey, R., and Avants, J. K. 1978. Separation and characterization of endo-polygalacturonase and exo-polygalacturonase from peaches. *Plant Physiol.* 52:252–256.

Seyderhelm, I., Boguslawski, S., Michaelis, G., and Knorr, D. 1996. Pressure induced inactivation of selected food enzymes. *J. Food Sci.* 61:308–310.

Verlinden, B. E., and De Baerdemaeker, J. 1997. Modeling low temperature blanched carrot firmness based on heat induced processes and enzyme activity. *J. Food Sci.* 62:213–218, 229.

Verlinden, B. E., De Barsy, T., De Baerdemaeker, S., and Deltour, R. J. 1996. Modeling the mechanical and histological properties of carrot tissue during cooking in relation to texture and cell wall changes. *J. Texture Stud.* 27:15–28.

Vilariño, C., Del Giorgio, J. F., Hours, R. A., and Cascone, O. 1993. Spectrophotometric method for fungal pectinesterase activity determination. *Lebensm.-Wiss. U.-Technol.* 26:107–110.

CHAPTER 5

Viscoelastic Properties of Egg Gels Formed with High Hydrostatic Pressure

ALBERT IBARZ
ELBA SANGRONIS
LI MA
GUSTAVO V. BARBOSA-CÁNOVAS
BARRY G. SWANSON

INTRODUCTION

HEN eggs and their derivatives are considered a basic food due to their high nutrient and protein content. The most important parts of eggs are the whites and yolks, with the latter formed by a globular protein solution consisting of the ovomucine fiber, a phosphoglucoprotein, made of cisteine and metionine. Other significant proteins are ovotransferrine, ovomucine, and lysozime (Stadelman and Cotterill, 1986). The egg yolk is a dispersion of different kind of particles suspended in a proteic solution of its most important elements: fosfovitine, lipoviteline, lipovitelenine, liviteline, and ovoviteline. To market the egg and its parts, a pasteurization treatment is necessary to avoid microbial contamination, but this is difficult as egg proteins are thermosensitive and denature with intense pasteurization. To avoid the negative effect produced by thermal treatments, the high-pressure process is proposed as an alternative.

When the egg white is treated at pressures greater than 5,000 kg/cm^2, it coagulates and forms a gel (Bridgman, 1914). Okamoto et al. (1990) studied the effect of high pressure on hen eggs and compared the texture of the gels obtained with those treated with heat. Egg yolk under a pressure of 4,000 kg/cm^2 for 30 min at 25°C forms a gel, but it takes 5,000 kg/cm^2 for egg white to coagulate partially and 6,000 kg/cm^2 for gelation to be complete. Egg white gels formed with pressure possess their natural flavor and have no loss of vitamins and amino acids, in addition to being more easily digested than gels formed with heat. The pressure-induced gels also retain their original yolk color and are soft, lustrous, and adhesive compared to heat-induced gels (Okamoto et al., 1990).

57

Boiled eggs often have a sulfur flavor and contain lysinoalanine produced during cooking, which inhibits the activation of proteolytic enzymes in the intestine and, therefore, reduces the availability of amino acids to the human body. The gels formed under pressure do not present this problem.

All egg parts have a large amount of protein, and it is assumed that gel formation causes the denaturation of these proteins. Grant et al. (1941) verified that coagulation by high pressure is associated with protein denaturation, and they showed that SH groups were reactive after such treatment. It is also believed that the denaturation of proteins by pressure is due to rearrangement and/or destruction of non-covalent bonds, such as hydrogen bonding, hydrophobic interactions, and ionic bonding contributing to tertiary structure, though covalent bonds are not affected.

Various methods can be used to characterize gels, but one of the most common is based on evaluation of their textural characteristics (Okamoto et al., 1990). Linear viscoelastic measurements are also particularly valuable in studying the viscoelastic behavior of different types of gels (Ma and Barbosa-Cánovas, 1995a, 1995b; Ma et al., 1996; Shoemaker et al., 1987).

The purpose of this work was to study the effect of high hydrostatic pressure treatment on gelation of egg yolk, egg white, and whole egg and to compare the viscoelastic characteristics of gels against the gels formed by heating under boiling water.

MATERIAL AND METHODS

SAMPLE PREPARATION

Egg samples (egg yolk, egg white, and whole egg) were obtained from fresh hens' eggs purchased in a local supermarket in Pullman, WA. The egg white chalazas had been removed previously, and the samples were stirred using a Hamilton Beach mixer, model 232B (Hamilton Beach/Proctor-Silex, Inc., Washington). Foams produced during the procedure were also removed, and the samples were then placed inside modified plastic syringes (2.5 cm in diameter and 5 cm in height).

HIGH HYDROSTATIC PRESSURE TREATMENT

The syringes containing the different samples were sealed in plastic bags filled with water and were treated with high hydrostatic pressure using an EPSI high hydrostatic press (Engineered Pressure Systems, Inc., Andover, MA) with a cylindrical pressure chamber (height:0.25 m, diameter: 0.1 m). A 5% Mobil Hydrasol 78 water solution was used as the pressure medium. The pressures applied were 410, 480, 550, 620, 650, and 690 MPa, and treatment times were 1, 3, 5, 10, 15, 20, and 30 min, but with the equipment used it generally took 5 min or less to reach the desired pressure (Figure 5.1).

HEAT TREATMENT

The syringes containing the different samples as described above were heated for 6, 8, and 10 minutes in a boiling water bath, then were cooled to room temperature, and then were cooled and kept at 4°C in a refrigerator.

RHEOLOGICAL MEASUREMENTS

Oscillatory shear experiments were performed in a control stress-control shear Physica Rheometer (Physica USA, Inc., Spring, TX) using parallel plate geometry (MP 30, 25 mm in diameter). The gel samples were allowed to rest for 5 min after loading to allow relaxation and temperature equilibration and were exposed to dynamic oscillatory tests to determine the viscoelastic parameters of the storage modulus (G') and loss modulus (G''). Frequency sweep tests were also performed using a selected oscillatory range of 0.63 to 63 rad/s (0.1 to 10 Hz) for a shear stress of 500 Pa, and the amplitude sweep tests utilized a fixed frequency of 6.3 rd/s (1 Hz). The sample was kept at room temperature (20°C) for all tests, and data reported are the average of three replicates.

RESULTS AND DISCUSSION

For the egg yolk samples, gels were formed at pressures of 410 MPa, while for whole egg and egg white samples, the pressures were 650 MPa (Table 5.1). It can be observed that egg yolk gels were formed for 1 min with 650 to 690 MPa, but 3, 5, and 20 min were necessary with 550, 480, and 410 MPa, respectively. Egg white formed hard gels at 3, 15, and 30 min with 690, 650, and 620 MPa, respectively. Hard gels for whole egg were obtained at 1, 3, and 10 min with 690, 650, and 620 MPa, respectively.

EGG YOLK

Figure 5.2 shows an amplitude sweep test for boiled egg yolk samples at a fixed frequency of 1 Hz. The storage and loss moduli decreased slightly with applied shear stress over the experimental range, and it can be observed that the values for both moduli also decrease when the heating time decreases. The elastic and loss moduli of the boiled egg yolk samples with different heating times showed similar trends with respect to shear frequency (Figure

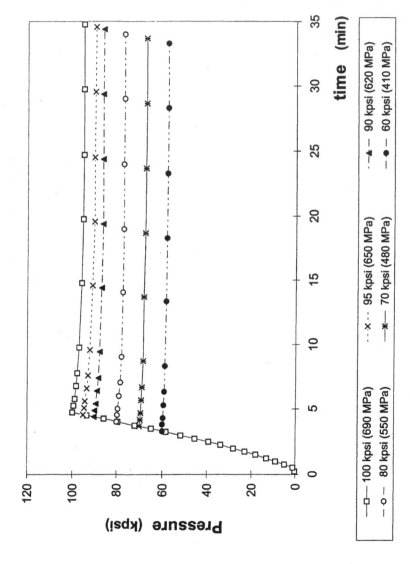

Figure 5.1 Pressure profiles in the chamber operation.

TABLE 5.1. Conditions for Gelation of Samples Under High Hydrostatic Pressure.*

Egg Sample	Pressure (Kpsi)	Time (min)						
		1	3	5	10	15	20	30
Yolk	100	+	+	+	+	+	+	+
	90	+	+	+	+	+	+	+
	80	o	+	+	+	+	+	+
	70	o	o	+	+	+	+	+
	60	o	o	o	o	o	+	+
White	100	o	+	+	+	+	+	+
	95	o	o	o	o	+	+	+
	90	o	o	o	o	o	o	+
Whole	100	+	+	+	+	+	+	+
	95	o	+	+	+	+	+	+
	90	o	o	o	+	+	+	+

* +, complete gelation; o, incomplete gelation.

5.3), but those heated for 6 and 8 min had lower values than those heated for 10 min.

Figure 5.4 shows the variation in storage and loss moduli for an amplitude sweep test on gels formed with pressure for 5 min compared to those gels obtained after boiling in water for 10 min. It can be observed that both moduli slightly decreased while shear stress increased, and though the storage and loss moduli values for boiled samples were greater than for the pressurized samples, they increased when the treatment pressure was increased. In the frequency sweep tests (Figure 5.5), the storage and loss moduli exhibited a slight increase when the frequency increased, which was similar to previous tests, but the values for gels formed with 620 and 690 MPa were also very close.

Figure 5.6 presents the variation in storage and loss moduli for an amplitude sweep test with gels formed with pressure for 30 min. The storage and loss moduli values for the boiled samples were found to be greater than those for the pressurized samples. Both moduli, G' and G'', exhibited similar values for all pressures unless formed under 410 Mpa, which made the latter's values appreciably less. The frequency sweep tests (Figure 5.7) revealed that the values for both moduli increased when the pressure was increased until a maximum value for gels formed with 550 MPa was reached. The gels tend to collapse at high pressure and long treatment times and depending on the denaturation of proteins.

In all samples, the storage moduli were greater than the loss moduli, and both were affected by shear stress in a similar fashion.

EGG WHITE

Fresh egg white turned opaque with partial coagulation when it was heated

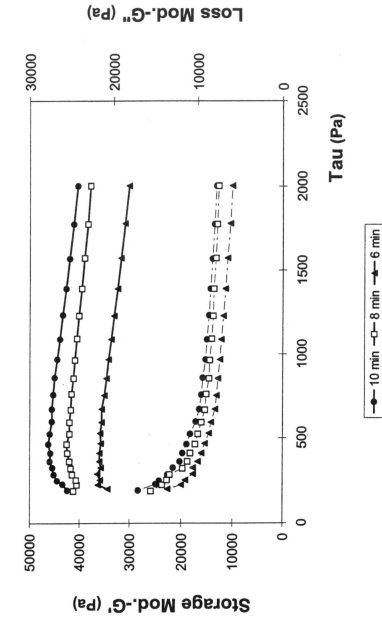

Figure 5.2 Amplitude sweep profile of boiled egg yolk. Continuous line G' and discontinuous line G''.

62

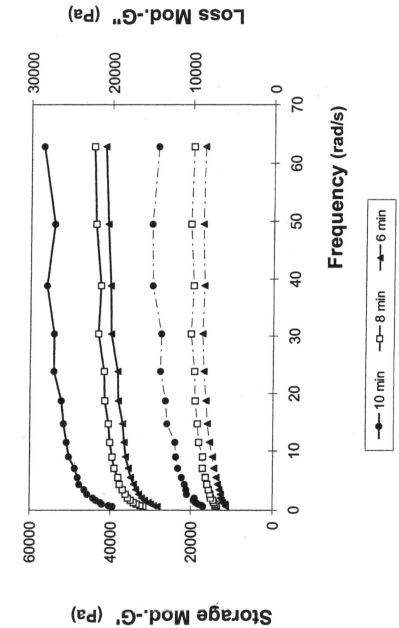

Figure 5.3 Frequency sweep profile of boiled egg yolk. Continuous line G' and discontinuous line G''.

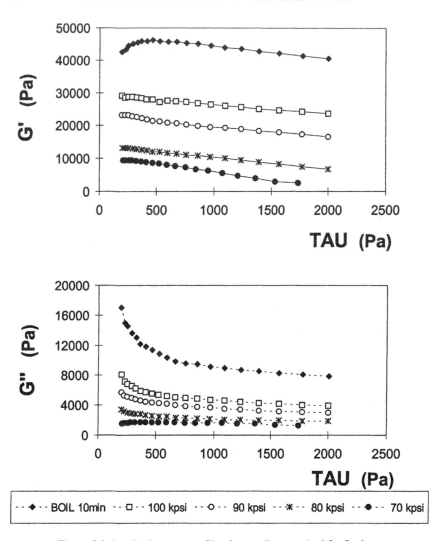

Figure 5.4 Amplitude sweep profile of egg yolk pressurized for 5 min.

in boiling water for 6 min and formed hard gels after 8 min. To conduct the HHP treatment, it was necessary to exert a pressure of 620 MPa to obtain hard gels. The egg white gels were less consistent than the egg yolk gels, possibly due to their lower protein content. The egg white gels were also unable to undergo the frequency sweep tests, because they were broken in the experimental test. The amplitude sweep tests revealed greater storage and loss moduli values for the boiled gels than for the pressurized gels, but a shear stress of 365 Pa was required to compare the consistency of the different gels

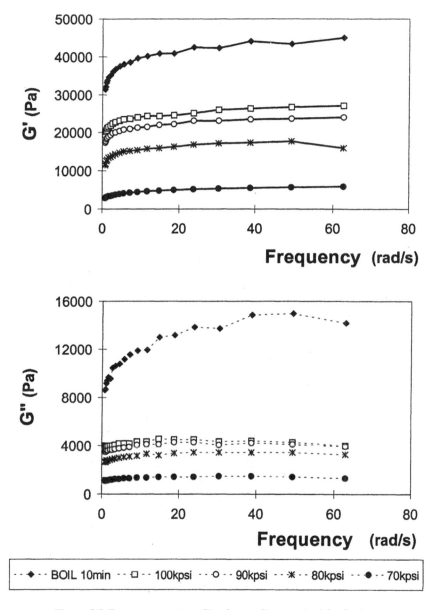

Figure 5.5 Frequency sweep profile of egg yolk pressurized for 5 min.

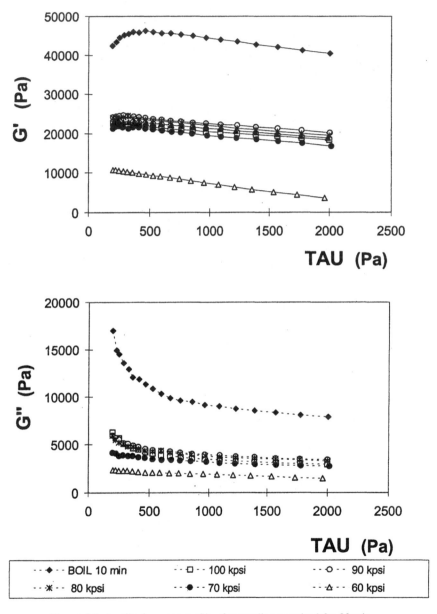

Figure 5.6 Amplitude sweep profile of egg yolk pressurized for 30 min.

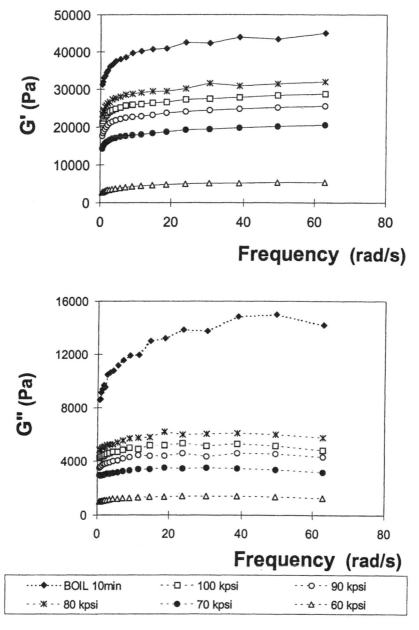

Figure 5.7 Frequency sweep profile of yolk pressurized for 30 min.

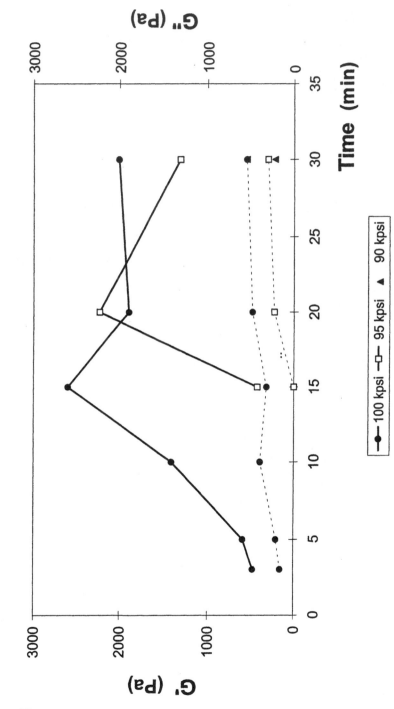

Figure 5.8 Variation of G' and G'' with a process time for egg white, at $\tau = 365$ Pa, in an amplitude sweep test. Continuous line G' and discontinuous line G''.

(Figure 5.8). The storage modulus values were 8.52 and 12.09 KPa, and the loss modulus values were 1.63 and 2.37 KPa for gels boiled for 8 and 10 min, respectively. For pressure-induced gels, the loss modulus increased when pressure and treatment times increased, while the storage modulus increased with short applications of pressure, but decreased with extensive ones.

WHOLE EGG

The whole egg gels showed similar behavior to that of the egg yolk and egg white on their amplitude sweep tests, although the storage and loss moduli values were similar to the egg white; their consistency was compared based on a shear stress of 595 Pa (Figure 5.9). For the same shear stress, the storage modulus values were 17.0, 25.6, and 26.8 KPa for gels boiled for 6, 8, and 10 min, respectively, while loss modulus values were 3.21, 5.53, and 5.77 KPa. The frequency sweep tests showed the storage and loss moduli values as similar to the samples boiled for 8 and 10 min, while those boiled for 6 min had values in between these and the pressurized gels. In order to compare the different gels, a variation of both G' and G'' based on time were plotted for the different pressures studied at a fixed frequency of 30.4 rad/s (Figure 5.10). The corresponding storage modulus values were 12.01, 20.99, and 21.53 KPa for gels boiled at 6, 8, and 10 min, respectively, and the loss modulus values were 2.59, 4.79, and 5.12. The behavior of the pressure-induced gels on amplitude and frequency sweep tests were similar, with the storage and loss moduli increasing when the time increased for gels formed at 620 and 650 MPa; however, for gels formed with 690 MPa, this trend was truncated for gels that took 15 min to form and where the G' and G'' values were less than those of gels formed with 650 MPa. The most consistent gels were therefore obtained with a pressure of 650 MPa over time periods greater than 15 min.

The gelation mechanism is different for gels formed by heat and pressure. All samples are solutions with high protein content stabilized by hydrophobic interactions, hydrogen bridges, electrostatic interactions, and covalent bonds, including disulfide bridges. Heat induces protein denaturation by destruction or formation of covalent bonds (Okamoto et al., 1990). However, pressure does not affect these bonds, but decreases the reaction volume and produces protein denaturation because hydrophobic interactions increase and the quaternary structure of proteins are affected (Tauscher, 1995). Likewise, it could be that a rearrangement of water molecules surrounding amino acids and a denaturation process are favored. It was observed that egg yolk gels were stronger than whole egg gels, and whole egg gels were stronger than egg white gels. This is because the protein content is the highest for egg yolk gels and the lowest egg white gels. For short treatment times, egg yolk gels were stronger when the pressure was increased, but gels obtained for 30 min in the

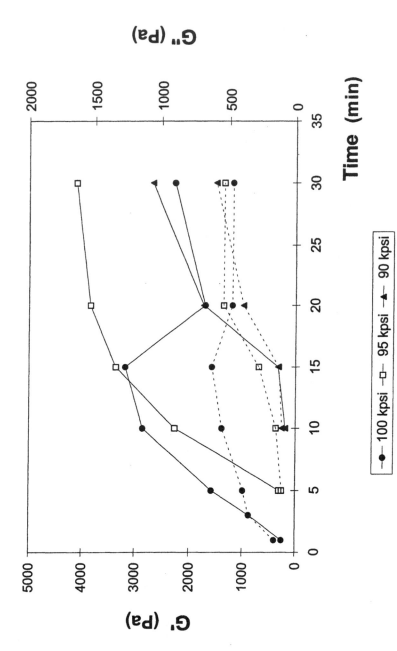

Figure 5.9 *Variation of G' and G'' with a process time for whole egg, at $\tau = 595$ Pa, in an amplitude sweep test. Continuous line G' and discontinuous line G''.*

70

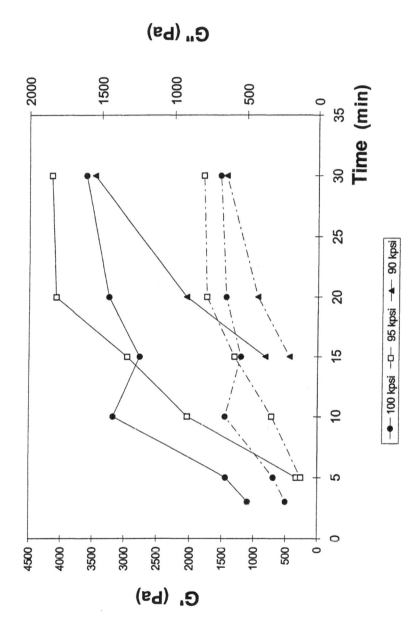

Figure 5.10 Variation of G' and G'' with processed time for whole egg, at $N = 30.4$ rad/s, in a frequency sweep test. Continuous line G' and discontinuous line G''.

71

range 480 to 690 MPa seemed similar. It is likely that long treatment times, at this pressure range, promote the breakage of the structural network of the gel, making no difference in the pressure level. The same trend and explanation could be given for egg white and whole egg gels.

ACKNOWLEDGEMENT

Author A. Ibarz is grateful to Ministerio de Educación y Ciencia (Spain) for the financial support (PR95-131) that allowed his visit to Washington State University.

REFERENCES

Bridgman, P. W. 1914. The coagulation of albumen by pressure. *J. Biol. Chem.* 19:511–512.

Grant, E. R., Dow, R. B., and Franks, W. R. 1941. Denaturation of egg albumin by pressure. *Science.* 94:616.

Ma, L., and Barbosa-Cánovas, G. V. 1995a. Rheological characterization of mayonnaise. Part I. Slip velocity and yield stress of mayonnaise as a function of oil and xanthan concentration. *J. Food Eng.* 25:397–408.

Ma, L., and Barbosa-Cánovas, G. V. 1995b. Rheological characterization of mayonnaise. Part II. Flow and viscoelastic properties of mayonnaise as a function of oil and xanthan cencentration. *J. Food Eng.* 25:409–425.

Ma, L., Grove, A., and Barbosa-Cánovas, G. V. 1996. Rheological properties of surimi gels: Effects of setting and starch. *J. Food Sci.* 61(5):881–883, 889.

Okamoto, M., Kawamura, Y., and Hayashi, R. 1990. Application of high pressure to food processing: Textural comparison of pressure and heat-induced gels of food proteins. *Agric. Biol. Chem.* 54(1):183–189.

Shoemaker, C. F., Lewis, J. I., and Tamura, M. S. (1987). Instrumentation for rheological measurements of food. *Food Technology.* 41(3):80.

Stadelman, W. J., and Cotterill, O. J. 1986. *Egg Science and Technology.* 3rd ed. Wesport, CT: AVI Pub. Co.

Tauscher, B. 1995. Pasteurization of food by hydrostatic pressure: Chemical aspects. *Z. Lebensm. Unters Forsch.* 200:3–13.

CHAPTER 6

Browning of Apple Slices Treated with High Hydrostatic Pressure

ALBERT IBARZ
ELBA SANGRONIS
GUSTAVO V. BARBOSA-CÁNOVAS
BARRY G. SWANSON

INTRODUCTION

BROWNING reactions in foods lead to major economic losses during processing and storage. Until recently, both enzymatic and non-enzymatic browning of foods could be inhibited by application of sulfites. However, sulfites are associated with allergy reactions in some people with asthma. Therefore, the Food and Drug Administration (FDA) has limited sulfites to certain categories of food products (FDA, 1990). Food processors have turned to sulfite alternatives, generating considerable research activity in this area. Several promising new browning-inhibitor treatments have been examined with varying success.

Enzymatic browning in fruit results when monophenolic compounds are hydrolyzed to o-diphenols, and the latter are oxidized to o-quinones. The presence of atmospheric oxygen and polyphenoloxidase (PPO) are required for browning reactions (Vámos-Vigyázo, 1981; McEvilly et al., 1992). The quinones condense and react nonenzymatically with other phenolic compounds, amino acids, etc., producing pigments of indeterminate structure. A variety of phenolic compounds are oxidized by PPO. The most important substrates are catechins, ester 3,4-hydroxyphenylalanine (DOPA), cinnamic acid, and tyrosine (Saper, 1993). PPO activity is prevented by heating, excluding oxygen, lowering the pH, removing or transforming the substrate, or adding inhibitors of enzymatic browning (Whitaker and Lee, 1995).

The inhibitors of enzymatic browning most frequently used in the apple industry are ascorbic acid and its isomer erythorbic acid, due to their ability to reduce quinones. These compounds can be added to syrups or dipping

73

solutions (Saper, 1993; Luo and Barbosa-Cánovas, 1995). The activity of PPO is also inhibited at pH 2 or less. Thus, citric acid is used as an acidulant inhibiting the enzyme (Whitaker and Lee, 1995). Even though the mechanism is not clear, 4-hexylresorcinol is a very good inhibitor of enzymatic browning in shrimp, apples, and potatoes (Luo and Barbosa-Cánovas, 1995). Non-enzymatic browning via Maillard-type reactions is another important route of color formation in apples. The reaction occurs between amino acids and reduces sugars present in the fruit, causing undesirable color, odor, and flavor changes. Phenolic compounds can also undergo oxidation, producing brown color (Saper, 1993). A simple and effective way to study the kinetics of deterioration of fruit and juices is through color measurements. Brown color development assumes zero-order reaction kinetics (Beveridge and Harrison, 1984; Toribio and Lozano, 1984; Monsalve-González et al., 1993).

High pressure is used to preserve foods because many microorganisms are sensitive to pressure (Sale et al., 1970). Because high pressure can modify protein structure, the food enzymes also change their behavior. However, with PPO, the results of using high pressure are not clear. Asaka and Hayashi (1991) found that high pressure increased PPO activity in pears, whereas other authors (Knorr, 1993) have reported PPO inhibition in potatoes. According to Gomes and Ledward (1996) and Eshtiaghi et al. (1994), the enzymes present in crude extracts obtained from the food are differently affected by pressure than commercially produced enzymes.

The objectives of this research were (1) to evaluate the browning development on apple slices treated with high pressure combined with some browning inhibitors and (2) to develop an empirical equation that describes the changes of color as a function of pressure, time, and type of inhibitory solution.

MATERIALS AND METHODS

SAMPLES

Golden Delicious and Granny Smith apples were purchased in a local food store in Pullman, WA. Apples slices were obtained by cutting the fruit along the stem-calyx axis with a stainless-steel hand slicer.

HIGH HYDROSTATIC PRESSURE TREATMENTS

Apple slices were placed into plastic bags (9 cm × 5 cm) adding about 50 ml of distilled water or inhibiting browning solutions. Ascorbic acid, citric acid, and 4-hexylresorcinol (4-HR) were the compounds used to inhibit browning in apple slices. Each solution contained 50 ppm of each inhibitor. The plastic bags were heat sealed and treated with high hydrostatic pressure. The pressures applied were 410, 480, 550, 620, and 690 MPa, and treatment times were 5, 10, 20, and 30 min. An EPSI high hydrostatic press was used (Engineered Pressure Systems, Inc., Andover, MA) with a cylindrical pressure chamber (height: 0.25 m, diameter: 0.1 m). A 5% Mobil Hydrasol 78 water solution was used as the pressure medium. The pressure profiles for the assayed treatments were determined as presented in Figure 6.1, which shows the needed time to reach required pressures. Light changes in pressure were observed during the treatment period. All tests were conducted in triplicate.

COLOR MEASUREMENTS

Treated apple slices were drained, rinsed with running water, and dried with absorbent paper. Slices were extended on trays in contact with surrounding air at room temperature (20°C to 25°C). Color changes of untreated and treated apple slices were evaluated during eight hours. The chromatic parameters L^*, a^*, and b^* were measured with a Minolta CM-2002 Spectrophotometer (Minolta Camera Co., Ltd., Osaka, Japan) in 60-min intervals.

STATISTICAL ANALYSIS

Means and standard deviation of measurements were calculated. Analysis of variance, LSD, and modeling were conducted using STATGRAPHIC 7.0 (Statgraphics, 1993).

RESULTS AND DISCUSSION

Browning of the apple slices was measured by L^* (lightness), b^* (yellow blue), and a^* (green red). A decrease in L^* and an increase in a^* indicates browning (Saper and Douglas, 1987). Graphs were obtained plotting L^* versus time, at each processing pressure and processing time. Changes of L^* color parameters for Granny Smith and Golden Delicious varieties had the same behavior. The initial L^* values of high pressure-treated apple slices were lower than those of untreated apple slices. For all experimental pressures, the initial values varied from 55 to 62. Because the trend for all treatments was similar, only the extreme treatment conditions are shown (Figures 6.2, and 6.3). During the eight hours of experimentation, there were no important changes in L^* values of apple slices immersed in different inhibitory solutions and treated at 410 MPa for 0, 1, 5, 10, 20, and 30 min [Figure 6.2(a)]. The apple slices immersed in 4-HR gave the highest L^* value. The apple slices immersed in different inhibitory solutions and treated at 690 MPa for 0, 1, 5, 10, 20, and 30 min [Figure 6.2(b)] gave similar initial L^* values as the apple slices treated at 410 MPa. During the period of experimentation, the L^* variation was not

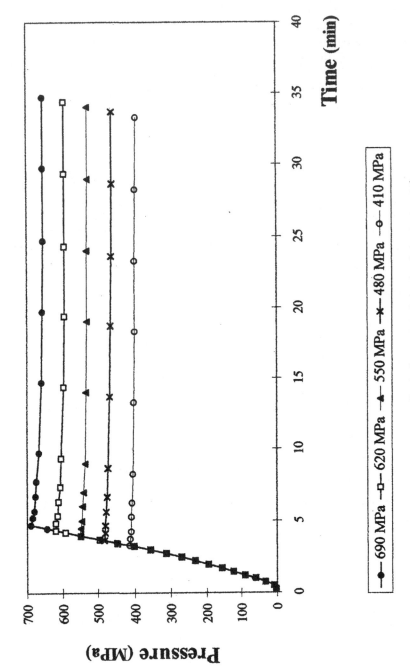

Figure 6.1 Pressure profiles in the pressurized chamber during treatment time.

Legend: —●— 690 MPa —□— 620 MPa —▲— 550 MPa —✳— 480 MPa —○— 410 MPa

a) Treated at 410 MPa for 1 min

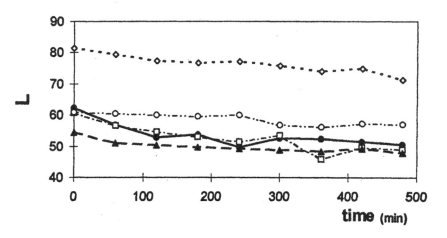

b) Treated at 690 MPa for 30 min

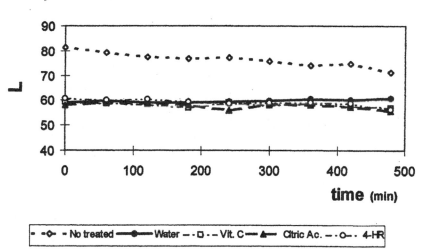

- · -◇- · - No treated ——●—— Water – · -□- · – Vit. C —— ▲ —— Citric Ac. – · -○- · - 4-HR

Figure 6.2 Evolution of L^* parameter for Granny Smith apple slices.

a) Treated at 410 MPa for 1 min

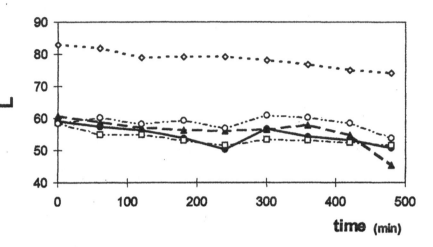

b) Treated at 690 MPa for 30 min

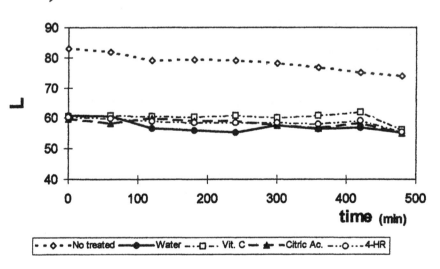

- - ◇ - - No treated ——●——Water - - -□- - . Vit. C — ▲- —Citric Ac. - - -○- - - 4-HR

Figure 6.3 Evolution of L^* parameter for Golden Delicious apple slices.

TABLE 6.1. Parameters of Equation (2) for Non-Treated Apple Slices.

Apple Variety	a^*_∞	$(a^*_\infty - a^*_0)$	$k_x\,10^3$ (min^{-1})	r^2
Golden Delicious	11.85	8.75	5.71	0.977
Granny Smith	14.66	12.96	1.70	0.983

TABLE 6.2. Evolution of a^* Parameter for Golden Delicious Apple Slices Dipped in Water.

Pressure (MPa)	Time (min)	a^*_∞	$(a^*_\infty - a^*_0)$	$k_x\,10^3$ (min^{-1})	r^2
410	1	10.99	8.32	6.004	0.988
	5	12.92	10.01	3.459	0.984
	10	9.92	7.54	6.631	0.992
	20	9.16	6.26	5.595	0.970
	30	10.45	7.67	3.870	0.977
480	1	14.26	10.94	7.710	0.994
	5	14.61	12.08	4.358	0.993
	10	14.80	13.18	4.759	0.993
	20	24.33	23.49	1.586	0.991
	30	10.68	9.95	2.845	0.976
550	1	12.99	9.60	2.206	0.953
	5	14.56	12.05	1.742	0.988
	10	9.41	7.33	4.003	0.995
	20	8.17	6.36	4.587	0.990
	30	4.79	4.19	1.694	0.993
620	1	10.45	6.97	5.682	0.983
	5	11.11	8.57	3.562	0.997
	10	10.73	7.94	4.110	0.998
	20	11.14	9.20	2.214	0.989
	30	9.32	8.34	2.155	0.991
690	1	13.55	10.84	3.120	0.988
	5	11.19	9.57	2.822	0.989
	10	11.96	10.73	1.416	0.997
	20	7.03	5.90	2.207	0.991
	30	8.05	6.87	4.460	0.988

TABLE 6.3. Evolution of a^* Parameter for Golden Delicious Apple Slices Dipped in Ascorbic Acid Solution.

Pressure (MPa)	Time (min)	a^*_∞	$(a^*_\infty - a^*_0)$	$k_x\,10^3$ (min^{-1})	r^2
410	1	10.01	7.29	9.966	0.988
	5	14.19	11.41	2.624	0.975
	10	10.64	7.82	5.178	0.991
	20	13.83	11.26	3.736	0.987
	30	15.94	13.41	1.762	0.988
480	1	14.57	10.98	5.987	0.991
	5	24.47	21.29	1.407	0.986
	10	13.51	10.89	4.043	0.988
	20	14.55	12.55	3.887	0.985
	30	7.57	6.78	5.570	0.962
550	1	11.15	8.55	2.518	0.986
	5	9.96	7.49	5.792	0.987
	10	9.35	7.37	3.425	0.986
	20	19.80	17.83	0.911	0.969
	30	10.16	9.15	1.039	0.997
620	1	12.22	8.55	3.875	0.984
	5	10.01	7.63	4.019	0.992
	10	13.73	11.02	1.511	0.992
	20	11.67	9.65	1.442	0.983
	30	5.77	4.83	3.287	0.996
690	1	19.19	17.57	1.423	0.995
	5	11.07	9.01	1.956	0.986
	10	10.57	9.91	0.942	0.963
	20	6.52	5.14	1.420	0.983
	30	10.09	8.13	2.038	0.991

significant. The high hydrostatic pressure-treated apple slices showed more browning than untreated apple slices. The b^* parameter of treated apple slices was not included in the browning study because its trend was not clearly defined. For the Granny Smith variety, treatments at 410 MPa increased a^* values for apple slices immersed in water, ascorbic acid, and citric acid. The apple slices immersed in 4-HR showed a^* values lower than untreated apple slices. Browning inhibition of Granny Smith slices was observed using 4-HR as inhibitory solution at all processing pressures and times. The Golden Delicious variety behaved differently. The initial a^* values for untreated apple slices were higher than treated slices. The variation with time for untreated

TABLE 6.4. Evolution of *a** Parameter for Golden Delicious Apple Slices Dipped in Citric Acid Solution.

Pressure (MPa)	Time (min)	a^*_∞	$(a^*_\infty - a^*_0)$	$k_x\ 10^3$ (min^{-1})	r^2
410	1	10.39	7.29	9.402	0.984
	5	12.58	8.71	7.355	0.992
	10	9.96	7.68	7.276	0.968
	20	10.54	7.89	4.123	0.969
	30	12.30	9.85	3.746	0.989
480	1	15.20	11.48	5.196	0.989
	5	14.52	10.79	7.961	0.992
	10	17.59	15.06	3.168	0.996
	20	12.97	10.79	5.037	0.975
	30	7.57	6.48	6.173	0.986
550	1	12.05	9.33	2.742	0.929
	5	15.23	13.15	1.124	0.962
	10	8.83	6.70	4.342	0.995
	20	10.51	9.05	2.660	0.987
	30	9.05	7.78	1.676	0.980
620	1	10.83	7.14	6.071	0.980
	5	14.29	10.83	3.259	0.997
	10	9.53	6.90	4.913	0.997
	20	9.65	7.26	2.735	0.979
	30	7.03	5.97	4.047	0.995
690	1	15.97	13.80	1.526	0.997
	5	10.44	8.31	2.921	0.993
	10	9.21	7.62	1.931	0.992
	20	10.35	9.25	0.929	0.965
	30	12.62	10.17	1.357	0.980

apple slices was lower than for treated apple slices. At the end of the experiment, *a** values for untreated apple slices were only higher than those of apple slices immersed in 4-HR. Browning inhibition of Golden Delicious slices was observed using 4-HR.

The results demonstrate that the most important variation of apple slices color was observed in the *a** parameter. Therefore, *a** variation as a function of exposure time was investigated. For different treatment conditions, the initial a^*_0 value evolved into a new value named a^*_∞. Then, the experimental data were fitted using a least square nonlinear parameter algorithm. Therefore, we can suppose that for a given time, *a** acquires a value calculated by this equation:

TABLE 6.5. Evolution of a^* Parameter for Golden Delicious Apple Slices Dipped in 4-HR Solution.

Pressure (MPa)	Time (min)	a^*_∞	$(a^*_\infty - a^*_0)$	k_x 10^3 (min^{-1})	r^2
410	1	9.76	7.64	1.060	0.913
	5	3.45	1.69	5.221	0.947
	10	4.63	3.06	9.341	0.960
	20	3.25	1.69	2.740	0.960
	30	3.66	2.06	4.221	0.977
480	1	5.86	3.73	3.645	0.976
	5	7.18	5.00	2.911	0.976
	10	7.14	6.09	2.509	0.976
	20	4.11	3.26	7.541	0.975
	30	1.98	1.56	9.748	0.983
550	1	5.54	3.81	1.121	0.965
	5	4.05	2.39	4.306	0.951
	10	3.39	2.47	3.179	0.973
	20	2.57	1.41	5.674	0.958
	30	4.58	3.50	0.666	0.913
620	1	5.03	2.68	2.875	0.990
	5	3.59	1.77	6.168	0.975
	10	8.05	6.36	0.795	0.945
	20	3.11	1.58	4.530	0.907
	30	1.49	0.94	5.707	0.826
690	1	8.71	7.24	0.733	0.825
	5	2.74	1.12	1.933	0.916
	10	2.09	1.33	3.639	0.911
	20	3.16	1.96	0.975	0.918
	30	4.20	3.14	0.625	0.761

$$a^* = a^*_\infty - (a^*_\infty - a^*_0)\lambda \qquad (1)$$

where λ is a time-dependent parameter that is equal to 1 when t = zero. If exposure time is very long, λ is equal to zero. Then, the variation of a^* as a function of time turned out to be an exponential equation:

$$a^* = a^*_\infty - (a^*_\infty - a^*_0)\exp(-kt) \qquad (2)$$

where a^*_∞ and a^*_0 are fitting parameters and the difference $(a^*_\infty - a^*_0)$ is the maximum increase experimented by a^*; k is the reaction constant; and t is air exposure time. According to Equation (2), the change of a^* is a pseudo-

TABLE 6.6. Evolution of a^* Parameter for Granny Smith Apple Slices Dipped in Water.

Pressure (MPa)	Time (min)	a^*_∞	$(a^*_\infty - a^*_0)$	$k_x 10^3$ (min^{-1})	r^2
410	1	11.19	9.40	9.469	0.980
	5	12.84	9.97	8.624	0.987
	10	11.04	10.53	10.575	0.994
	20	7.95	8.07	6.274	0.955
	30	16.13	11.33	1.927	0.987
480	1	10.59	10.37	9.732	0.977
	5	12.62	11.39	8.348	0.993
	10	11.08	11.11	7.084	0.983
	20	13.12	12.27	4.373	0.989
	30	14.09	13.43	2.879	0.990
550	1	10.85	9.54	5.968	0.991
	5	7.60	7.82	4.739	0.989
	10	11.55	10.87	6.625	0.988
	20	9.08	9.15	3.690	0.970
	30	8.70	9.43	4.847	0.972
620	1	12.47	13.70	3.566	0.993
	5	9.38	10.88	3.855	0.987
	10	9.24	10.35	2.078	0.982
	20	7.39	9.06	1.189	0.974
	30	5.83	7.26	6.750	0.980
690	1	8.53	7.34	3.733	0.995
	5	15.44	14.58	1.755	0.962
	10	2.72	4.46	3.051	0.964
	20	4.68	6.10	6.727	0.965
	30	2.95	4.73	2.881	0.981

first-order reaction. The same type of equation was given for non-enzymatic browning in apple juice during storage (Toribio and Lozano, 1984) and in pear juice concentrate (Beveridge and Harrison, 1984). By substitution of experimental results in Equation (2), the variation of a^* was obtained (Table 6.1). Fitting parameters and estimated parameters were significant at $p < 0.05$. For the Golden Delicious variety (Tables 6.2–6.5), a^*_∞ was greater than a^*_∞ obtained for the Granny Smith variety (Tables 6.6–6.9). At any time, slices of Golden Delicious may be darker than slices of Granny Smith. On the contrary, the reaction constant (k) for browning of the Granny Smith variety is higher that the k value of Golden Delicious. That means that the rate of

TABLE 6.7. Evolution of a^* Parameter for Granny Smith Apple Slices Dipped in Ascorbic Acid Solution.

Pressure (MPa)	Time (min)	a^*_∞	$(a^*_\infty - a^*_0)$	$k_x\,10^3$ (min^{-1})	r^2
410	1	12.39	10.45	7.968	0.986
	5	9.75	7.34	8.666	0.988
	10	7.22	7.20	6.107	0.993
	20	8.13	7.84	13.003	0.983
	30	18.54	11.70	1.932	0.987
480	1	11.59	12.34	7.342	0.988
	5	13.36	12.08	5.701	0.997
	10	10.26	10.75	7.077	0.993
	20	10.56	9.87	6.847	0.989
	30	10.50	9.38	4.176	0.991
550	1	14.65	11.49	5.546	0.991
	5	10.92	9.33	5.292	0.976
	10	9.16	7.75	4.323	0.975
	20	7.61	6.89	6.047	0.976
	30	8.97	8.47	4.988	0.993
620	1	9.03	9.51	3.186	0.992
	5	9.16	10.12	1.720	0.987
	10	10.26	10.92	1.258	0.985
	20	9.96	11.79	1.117	0.987
	30	8.22	8.71	2.861	0.987
690	1	15.09	14.23	1.173	0.956
	5	22.99	21.97	1.144	0.973
	10	6.77	8.30	2.318	0.967
	20	4.47	4.59	6.076	0.972
	30	9.18	10.67	1.318	0.986

browning reaction in Granny Smith is higher than that of Golden Delicious. In both varieties, when 4-HR solution was used, the value of a^*_∞ was the lowest. This result indicates that apple slices immersed in 4-HR solution would have less color change as a function of time. The a^*_∞ values of slices immersed in citric acid and ascorbic acid solutions were higher than those slices immersed in other inhibitory solutions. The expression $(a^*_\infty - a^*_0)$ is a measure of color intensity following the same trend of a^*_∞. These results indicate that 4-HR was the best inhibitor for color changes followed by water, ascorbic acid, and citric acid. When 4-HR solution was used, the small k values indicated a browning reaction rate lower than for other dipping solutions.

TABLE 6.8. Evolution of a^* Parameter for Granny Smith Apple Slices Dipped in Citric Acid Solution.

Pressure (MPa)	Time (min)	a^*_∞	$(a^*_\infty - a^*_0)$	$k_x 10^3$ (min^{-1})	r^2
410	1	9.29	7.19	14.034	0.970
	5	12.52	9.72	6.382	0.975
	10	11.06	10.12	7.859	0.989
	20	9.90	8.44	11.708	0.970
	30	19.62	12.20	1.647	0.989
480	1	11.03	11.18	6.822	0.988
	5	11.33	11.46	6.268	0.996
	10	8.88	9.57	5.157	0.979
	20	9.46	8.10	4.559	0.963
	30	11.34	10.44	4.714	0.998
550	1	10.89	10.01	2.591	0.958
	5	11.91	11.27	4.026	0.985
	10	7.91	8.03	5.642	0.990
	20	5.37	6.62	3.714	0.972
	30	7.22	7.47	5.548	0.978
620	1	5.60	6.23	4.940	0.981
	5	6.95	7.33	3.027	0.991
	10	12.54	14.10	0.951	0.974
	20	4.14	6.00	4.762	0.993
	30	7.74	8.87	4.470	0.986
690	1	10.79	8.34	2.788	0.997
	5	17.29	16.42	1.306	0.985
	10	12.26	13.76	1.158	0.973
	20	13.07	14.39	1.519	0.987
	30	7.21	8.71	1.788	0.981

CONCLUSIONS

In this work, it was demonstrated that L^* values of apple slices previously immersed in inhibitor browning solution treated with high pressure were lower than for untreated slices. The L^* values did not change over time. Small a^* values were obtained when high pressure was applied to apple slices. For apple slices immersed in 4-HR solution, the increase of a^* was the lowest, and it was not affected by pressure. The same behavior was observed for varieties Granny Smith and Golden Delicious. Additionally, the a^* value increased as a function of time and followed a pseudo-first-order reaction.

TABLE 6.9. Evolution of a^* Parameter for Granny Smith Apple Slices Dipped in 4-HR Solution.

Pressure (MPa)	Time (min)	a^*_∞	$(a^*_\infty - a^*_0)$	$k_x\,10^3$ (min⁻¹)	r^2
410	1	11.02	11.27	1.284	0.991
	5	5.66	5.69	3.561	0.986
	10	5.54	5.69	3.102	0.976
	20	4.32	4.78	5.991	0.985
	30	4.36	3.09	3.139	0.990
480	1	2.26	4.90	3.221	0.985
	5	4.85	5.48	4.000	0.989
	10	2.90	4.07	3.674	0.974
	20	4.07	5.24	3.581	0.972
	30	4.62	4.20	3.653	0.962
550	1	5.91	4.89	1.801	0.929
	5	2.77	2.69	6.079	0.978
	10	1.83	2.62	6.827	0.977
	20	0.21	1.75	4.408	0.993
	30	5.32	6.61	1.908	0.987
620	1	3.48	5.36	3.868	0.993
	5	2.97	3.49	0.853	0.945
	10	0.16	1.94	3.152	0.981
	20	3.46	5.26	2.002	0.978
	30	4.31	6.92	1.317	0.990
690	1	3.56	3.81	2.463	0.985
	5	12.88	14.36	0.723	0.858
	10	2.68	4.91	0.853	0.890
	20	2.16	3.82	1.860	0.987
	30	5.98	7.69	1.253	0.981

The reaction constant, k, indicated that at a given time, the browning in Golden Delicious can be lower than in the Granny Smith variety.

ACKNOWLEDGEMENT

The author A. Ibarz is very grateful to CICYT [Ministerio de Educación y Ciencia (Spain) (PR95-13 1)] for the financial support received to spend three months at Washington State University.

REFERENCES

Asaka, M., and Hayashi, R. 1991. Activation of polyphenoloxidase in pear fruits by high-pressure treatment. *Agric. Biol. Chem.* 55:2439–2440.

Beveridge, T., and Harrison, J. E. 1984. Non-enzymatic browning in pear juice concentrate at elevated temperatures. *J. Food Sci.* 49:1335–1336, 1340.

Eshtiaghi, M. N., Stute, R., and Knorr, D. 1994. Effectiveness of high pressure pretreatment on dehydration rates, rehydration texture and color of rehydrated green beans, carrots and potatoes. *J. Food Sci.* 59:1168–1170.

FDA. 1990. Sulfiting agents: Affirmation of GRAS status. Food and Drug Admin. *Fed. Reg.* 57:51065–51084.

Gomes, M. R. A., and Ledward, D. A. 1996. Effect of high-pressure treatment on the activity of some polypheloxidase. *Food Chemistry* 56(1):1.

Knorr, D. 1993. Effects of high hydrostatic pressure on food safety and quality. *Food Technol.* 47 (6):156–161.

Luo, Y., and Barbosa-Cánovas, G. V. 1995. Inhibition of apple slice browning by 4-hexylresorcinol. Ch. 19, pp. 241–250. In *Enzymatic Browning and Its Prevention.* Lee, C. Y., and Whitaker, J. R., eds. American Chem. Soc. Washington, DC.

McEvilly, A. J., Iyengar, R., and Otwell S. W. 1992. Inhibition of enzymatic browning in foods and beverages. *Crit. Rev. Food Sci. Nutr.* 32 (3):253–273.

Monsalve-González, A., Barbosa-Cánovas, G. V., Cavalieri, R., McEvily, A. J., and Iyengar, R. 1993. Control of browning during storage of apple slices preserved by combined methods. 4-Hexylresorcinol as anti-browning agent. *J. of Food Sci.* 58 (4):797–800, 826.

Sale, A. J. H., Gould, G. W., and Hamilton, W. A. 1970. Inactivation of bacterial spores by hydrostatic pressure. *J. Gen. Microbiol.* 60:323–334.

Saper, G. M. 1993. Browning of food: Control by sulfites, antioxidants and other means. *Food Technol.* 47:75–84.

Saper, G. M., and Douglas, F. W., Jr. 1987. Measurement of enzymatic browning at cut surfaces and in juice of raw apple and pear fruits. *J. Food Sci.* 52:1258.

Statgraphics. 1993. Statistical Graphics System. Educational Institute Edition. Version 7.0. Rockville, Maryland.

Toribio, J. L., and Lozano, J. E. 1984. Nonenzymatic browning in apple juice concentrate during storage. *J. Food Sci.* 49:889–892.

Vámos-Vigyázo, L. 1981. Polypheloxidase oxidase and peroxidase in fruit and vegetables. *Crit. Rev. Food Sci. Nutr.* 15:49–127.

Whitaker, J., and Lee, C. Y. 1995. Recent advances in chemistry of enzymatic browning. Ch.2. In *Enzymatic Browning and Its Prevention.* Lee, C. Y., and Whitaker, J. R., eds. American Chem. Soc. Washington, DC.

CHAPTER 7

Evolution of Polyphenoloxidase (PPO) Activity in Apple Slices Treated with High Hydrostatic Pressure

ALBERT IBARZ
ELBA SANGRONIS
GUSTAVO V. BARBOSA-CÁNOVAS
BARRY G. SWANSON

INTRODUCTION

THERE are different forms of polyphenoloxidase (PPO) in plants and animals, but they all cause enzymatic browning in vegetal foods. This deteriorative process is undesirable in foods such as fruit juices and other fruit products. In order to avoid enzymatic browning in these foods, methods such as oxygen elimination, antioxidant addition (sulfur dioxide and ascorbic acid), and thermal treatments are used to inactivate PPO.

Recently, high hydrostatic pressure (HHP) processes have been used in food preservation because they are able to inactivate many microorganisms and minimize the loss of color and nutrients to a greater degree than those that occur during heat treatments. High pressure affects protein structure, and, therefore, enzymes are also affected. High pressure has been applied in the manufacture of jams from different fruits (Horie et al., 1981). Asaka and Hayashi (1991) found that Bartlett pear samples treated under pressures from 200 to 500 MPa grew dark quickly, so it would seem that Bartlett pear jams manufactured by HHP wouldn't be adequate for commercialization. The darkness of these samples is due to the action of polyphenoloxidase (o-biphenol: oxygen reductase, EC 1.10.3.1) (PPO) on natural substrates in the fruit. However, Knorr (1993) showed that pressure ranging from 100 to 400 MPa produced a decrease in potato PPO activity. Also, Eshtiaghi et al. (1994) showed that PPO from potatoes in a phosphate buffer solution (pH 7) was inactivated when pressurized under 900 MPa for 30 min at 45°C. Likewise, Seyderhelm et al. (1996) obtained similar results for their experiments with PPO inactivation. Gomes and Leward (1996) found that PPO activity extracted from mushrooms in a buffer solution at pH 7 decreased when pressurized in the 100 to 800 MPa range and was completely inactivated at the latter pressure. These authors also discovered that non-purified PPO from mushrooms pressurized under 400 MPa for 10 min showed a 40% increase in activity, while at 800 MPa, there was a 60% decrease. The PPO potato extract, from potato pressurized at 800 MPa for 10 min, showed a 40% residual activity compared to non-treated samples.

The action of PPO on phenolic substrates can be described by the Michaelis-Menten equation. For o-biphenol substrates, the product obtained is an o-quinone. The evolution of a substrate with the reaction time shows a maximum rate at the initial time for maximum substrate concentration. If the substrate concentration is less than the Michaelis-Menten constant (K_M), it is assumed to be a first-order kinetic, and the reaction rate can be expressed (Stauffer, 1989) as

$$r = \frac{dC_S}{dt} = -\frac{r_{MAX}}{K_M} \qquad (1)$$

The variation of substrate concentration with reaction time can be obtained by integrating this equation:

$$C_S = C_{S0} \exp(-kt) \qquad (2)$$

where C_S and C_{S0} are the substrate concentrations for a determined and initial time, respectively, and the constant k is

$$k = \frac{r_{MAX}}{K_M} \qquad (3)$$

The evolution of a substrate is a function of time, and the extinction molar coefficient is obtained for long reaction times, assuming that the concentration of a product is equal to its initial substrate concentration (Stauffer, 1989). For a given time, the substrate amount will be the initial concentration minus the amount reacted to the product.

The maximum rate value (r_{MAX}) at the initial time is obtained from the slope of the tangent of the substrate concentration time curve. According to Stauffer (1989), the r_{MAX} value can also be obtained by drawing the variation of product concentration/time versus time and extrapolating initial time.

The purpose of this work was to study the combined effect of HHP and antibrowning agents on the PPO activity of Granny Smith and Golden Delicious apple varieties and the possibility of controlling enzymatic browning.

MATERIALS AND METHODS

SAMPLES

Golden Delicious and Granny Smith apples were purchased in a local supermarket in Pullman, WA. The apples were cut in slices along their axial axis, using a stainless-steel manual cutter. The slices were next dipped in a 500 ppm ascorbic acid solution for two seconds and rinsed with water.

HIGH HYDROSTATIC PRESSURE TREATMENTS

Golden Delicious and Granny Smith apple slices were placed into plastic bags containing either water or 50 ppm solutions of ascorbic acid, citric acid, and 4-hexylresorcinol (4-HR). The plastic bags were then sealed and treated with HHP using an EPSI high hydrostatic press (Engineered Pressure Systems, Inc., Andover, MA) with a cylindrical pressure chamber (height: 0.25 m, diameter: 0.1 m) containing a 5% Mobil Hydrasol 78 water solution as the pressure medium. The pressures applied ranged from 138 to 690 MPa, and the treatment time was 5 min. All experiments were carried out in triplicate at room temperature (25 ± 0.5°C).

ENZYMATIC ACTIVITY MEASUREMENTS

Apple slices were washed with water and were homogenized with an equal volume of distilled water at 0°C. The homogenate was pressed manually, and the juice obtained was centrifuged at 6,000 rpm and 5°C for 20 min with a DuPont centrifuge model Sorvall RT 6000B. PPO was determined in 0.1 M sodium phosphate (pH 6.0) containing 50 mM catechol at 30°C as described by Moore and Flurkey (1990). The PPO activity was measured by absorbance at 410 nm using a spectrophotometer (Hewlet Packard 8452A). One unit of the activity was defined as a change in one absorbance unit at 410 nm per min per ml of juice.

RESULTS AND DISCUSSION

Figure 7.1 shows the pressure profiles for the assay treatments and the time needed to reach pressure. Slight changes in pressure were observed during the treatment period. All tests were conducted in triplicate.

Enzymatic activities measured at an absorbance of 410 nm per minute per ml of juice for the various treatment pressures are given in Figures 7.2 and 7.3 for the Granny Smith and Golden Delicious samples, respectively.

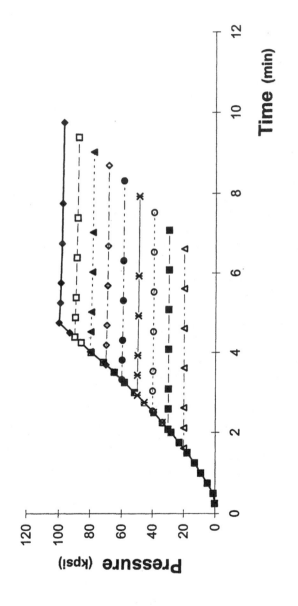

Figure 7.1 Pressure profiles in a pressurized chamber during treatment time.

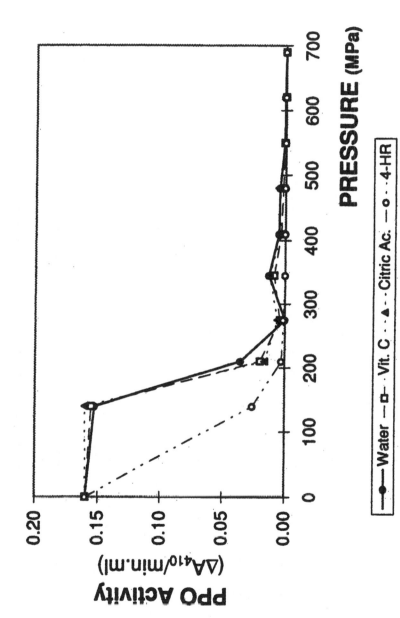

Figure 7.2 Evolution of Granny Smith PPO relative activity with applied pressure for different dipping solutions.

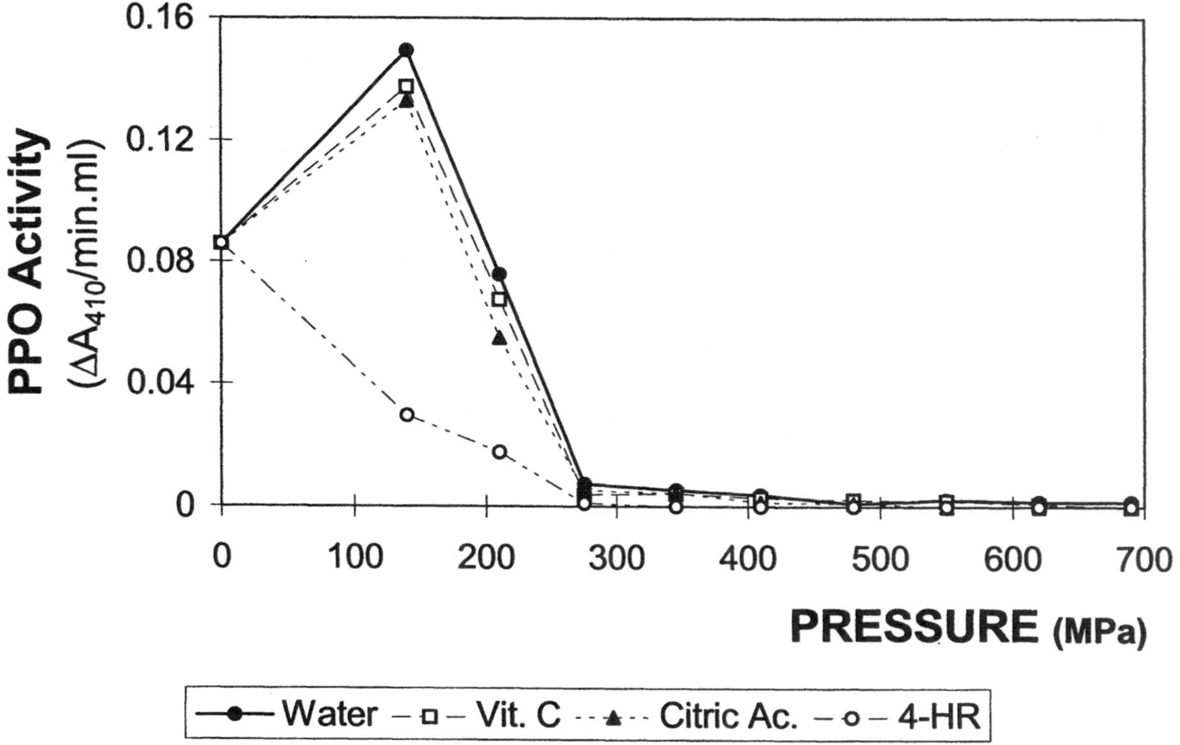

Figure 7.3 Evolution of Golden Delicious PPO relative activity with applied pressure for different dipping solutions.

94

TABLE 7.1. Michaelis-Menten Kinetic Constants for PPO Activity in Apples.

Apple Slices	$r_{MAX}\ 10^3$ (M/min)	$K_M\ 10^2$ (M)	$k10^2$ (min^{-1})
Golden Delicious	4.23	7.21	5.87
Granny Smith	8.30	7.41	11.20

For the Granny Smith samples at 138 MPa, enzymatic activity was observed to be analogous to the initial activity of the samples treated with water, ascorbic acid, and citric acid solutions. However, the enzymatic activities of samples treated in a solution of 4-HR were decreased considerably. At 210 MPa, in all cases, enzymatic activity was highly decreased. With treatment pressures above 275 MPa, residual enzymatic activity was also very low in all cases.

Enzymatic activity was observed to increase at 138 MPa in all Golden Delicious samples, except for those treated with a solution of 4-HR, where the activity decreased to one-third compared to the non-treated samples. When the treatment pressure was 210 MPa, all samples had decreased enzymatic activity. This was high for samples treated with 4-HR, while for those treated with water, the decrease was only a little higher than those not treated. As with the Granny Smith variety, from pressures of 275 MPa and up, residual enzymatic activity was very low in the Golden Delicious samples.

It is important to point out that the increases in enzymatic activity presented by Golden Delicious samples treated at 138 MPa were possibly due to breaks in cell walls that would cause the PPO to diffuse to the aqueous medium. At higher pressures, although cell breakage occurs (and therefore polyphenoloxidase liberation), there must be some denaturation of enzymes due to the effect of pressure. It is also important to emphasize that samples treated with a solution of 4-HR presented a much lower enzymatic activity. This is explained by the inhibiting effect of this product on PPO.

Kinetic constants of enzymatic activity for the various samples were also determined. Table 7.1 shows the results for the non-treated samples, where Granny Smith samples present a higher r_{MAX} value than Golden Delicious samples. The first-order global kinetic constant of Equation (2) was higher for the Granny Smith samples, which means that their enzymatic activity was also higher. For samples treated under pressure with the various aqueous media, the same parameters were obtained and then compared to the non-treated samples. Figures 7.4 and 7.5 show the maximum rate relations and first-order kinetic constants, respectively, for the Golden Delicious samples. Figures 7.6 and 7.7 show the same relationships for the Granny Smith samples. In all the figures, ordinate represents the parameter of samples not treated with pressure. For Golden Delicious samples, the r_{MAX} rate was observed to be higher in the samples treated with water than those not treated. When the

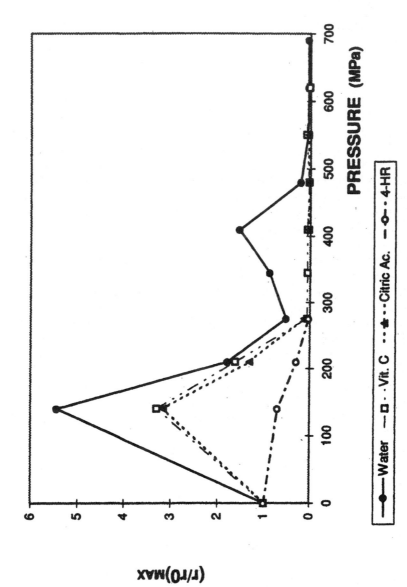

Figure 7.4 Evolution of the $(r/r_0)_{MAX}$ ratio with pressure for PPO in Golden Delicious apples.

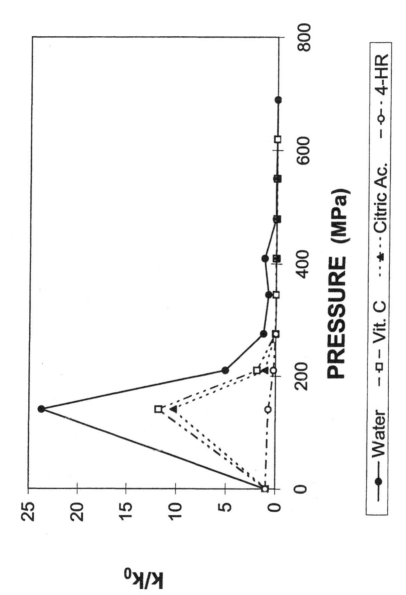

Figure 7.5 Evolution of the (k/k_0) ratio with pressure for PPO in Golden Delicious apples.

Legend: Water · · Vit. C · · · Citric Ac. · · -o- · 4-HR

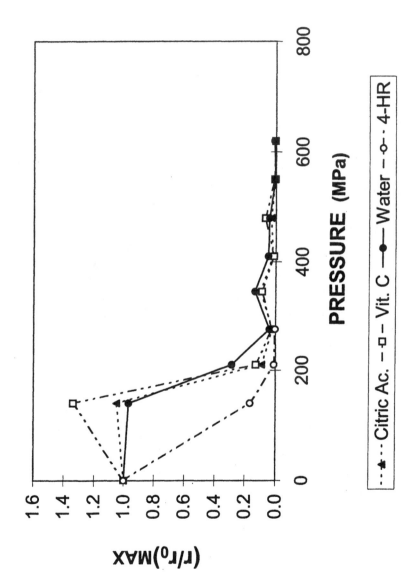

Figure 7.6 Evolution of the $(r/r_0)_{MAX}$ ratio with pressure for PPO in Granny Smith apples.

··▲··Citric Ac. --□--Vit. C —●—Water --○--4-HR

PRESSURE (MPa)

(r/r_0)MAX

98

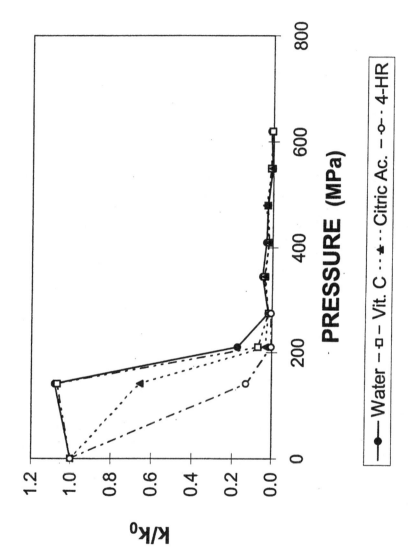

Figure 7.7 Evolution of the (k/k_0) ratio with pressure for PPO in Granny Smith apples.

treatment pressure was 138 MPa, the rate was lowest again at 275 MPa, then rose slightly with increasing pressure. At 410 MPa, its value was about 1.5 times higher; and at higher pressure, values obtained were very low. At pressures over 550 MPa, the r_{MAX} was no more than 1% of that of the non-treated sample.

For samples treated with ascorbic and citric acid solutions, an increase of the r_{MAX} was obtained at 138 MPa, which was nearly three times that of the non-treated sample. At 210 MPa, the values were similar, and at 275 MPa and higher, the r_{MAX} values for the pressurized samples were 3% lower than those of the non-treated samples. Those treated with 4-HR always showed lower r_{MAX} values compared to the untreated samples. At 138 MPa, the r_{MAX} values were about 70%, but at 275 MPa or higher, they were virtually zero.

Figure 7.6 shows that in Granny Smith samples, except for those treated at 138 MPa with citric and ascorbic acid solutions, the r_{MAX} was lower in samples treated under pressure; and with pressure higher than 275 MPa, maximum rates were below 5%.

The values of the first-order kinetic constants of the Golden Delicious samples treated with water, citric acid, and ascorbic acid solutions showed a great increase compared to untreated samples at 138 MPa, but decreased dramatically at pressures above 275 MPa. For samples treated with only water, this decrease started at 480 MPa. Samples treated with 4-HR always presented a stronger decrease, and the constant values were also consistently lower than the untreated samples. The Granny Smith samples at 138 MPa in citric acid and ascorbic acid solutions presented slightly higher constant values than the untreated samples. With all other treatment conditions, the constant values were always lower than those of the untreated samples, and from 275 MPa up, they did not pass the value of the untreated samples by more than 5%. It is important to point out that the samples treated with 4-HR always presented lower values than all other samples.

It can, thus, be concluded that to obtain apple derivatives of Granny Smith and Golden Delicious varieties by means of high pressure, it is advisable to work with pressures greater than 275 MPa. Lower pressures can be used if small quantities of 4-HR are introduced into the medium. Any of these operating methods would be adequate to prevent enzymatic browning caused by the action of PPO.

ACKNOWLEDGEMENT

The author A. Ibarz is very grateful to CIRIT [Generalitat de Catalunya (Spain)] for their financial support that permitted his visit to Washington State University.

REFERENCES

Asaka, M., and Hayashi, R. 1991. Activation of polyphenoloxidase in pear fruits by high pressure treatment. *Agric. Biol. Chem.* 55 (9):2439–2440.

Eshtiaghi, M. N., Stute, R., and Knorr, D. 1994. High-pressure and freezing pretreatment effects on drying, rehydration, texture and color of green beans, carrots and potatoes. *J. Food Sci.* 59:1168–1170.

Gomes, M. R. A., and Ledward, D. A. 1996. Effect of high-pressure treatment on the activity of some polyphenoloxidases. *Food Chem.* 56 (1):1–5.

Horie, Y. N., Kimura, K. I., and Ida, M. S. 1981. Jams treated at high pressure. US Patent 5,075,124.

Knorr, D. 1993. Effects of high-hydrostatic-pressure processes on food safety and quality. *Food Technol.* 47:156–161.

Moore, B. M., and Flurkey, W. H. 1990. Sodium dodecyl sulfate activation of a plant polyphenoloxidase. *J. Biol. Chem.* 265 (9):4982–4988.

Seyderhelm, I., Boguslawski, S., Michaelis, G., and Knorr, D. 1996. Pressure induced inactivation of selected food enzymes. *J. Food Sci.* 61 (2):308–310.

Stauffer, C. E. 1989. *Enzyme Assays for Food Scientists.* New York: AVI Book Publish., Van Nostrand Reinhold, pp. 10–36.

CHAPTER 8

The Role of Pressure and Time on the Asymptotic Modulus of Alaska Pollock and Pacific Whiting Surimi Gels

JATUPHONG VARITH
GUSTAVO V. BARBOSA-CÁNOVAS
BARRY G. SWANSON

INTRODUCTION

SEVERAL studies on high-pressure effects on food gels have been published (Hoover, 1993; Ohmiya et al., 1987; Okamoto et al., 1990; Van Camp et al., 1996). Okamoto et al. (1990) reported that a pressure treatment of 400 MPa formed a gel in egg yolk and 500 MPa resulted in a partially coagulated egg white. The hardness of high hydrostatic pressure (HHP)-induced egg white gels was only about one-sixth of the hardness of heat-induced egg white gels (Okamoto et al., 1990). Tauscher (1994) hypothesized that the effect of high pressure on protein denaturation is small when compared to the effect of temperature: 100 MPa may be required to reach a gel strength equivalent to a temperature increase of 10°C. A whey protein concentrate (WPC) gel formed during high-pressure treatment was reported by Van Camp and Huyghebaert (1995). Heat-induced WPC gels are firm and dry, while WPC gels produced by HHP are soft and cure surrounded by non-incorporated liquid after pressurization (Van Camp and Huyghebaert, 1995).

High-pressure treatments denature proteins and induce gels in surimi. Traditionally, the Japanese process to gel surimi-based products uses heat (i.e., steaming or frying). Many studies have investigated heat-induced surimi gel properties and surimi-based product manufacturing processes (Kamazawa et al., 1995; Bertak and Karahadian, 1995; Yoo and Lee, 1993; Kamath et al., 1992). However, there have been few reports on surimi processed by HHP (Lanier, 1996; Shoji et al., 1990). Shoji et al. (1990) mentioned that fish protein gels can be formed with pressures ranging from 206 to 517 MPa at ambient or lower temperatures. Lanier (1996) reported that HHP surimi gelation is most likely due to intermolecular hydrophobic associations, whereas heat gelation is due to disulfide bonding. The advantages of HHP surimi processing without heat are a natural appearance and flavor. The possibility of obtaining high-pressure pasteurized surimi gels is of great interest to the surimi industry. However, more studies of the functional properties of HHP surimi gels are needed to provide fundamental knowledge for the development of surimi-based products and processes.

One method for characterization of the functional properties of HHP surimi gels is by using the stress relaxation test. By definition, stress relaxation is the decay of stress with time when the material is suddenly deformed to a given constant strain (Mohsenin, 1986). The stress required to hold the strain constant is measured as a function of time. Traditionally, the Maxwell model has been a popular mechanical analog model used to explain the stress relaxation phenomenon. However, the Maxwell model is difficult to evaluate, because non-linear regression is required to fit the curve. Peleg (1979) proposed a simplified method to represent the stress relaxation curve using the residual force after the asymptotic period (no change in stress with respect to time) as a quantity index of internal fracture, which occurs during deformation of the food material. The relaxation curve is then simplified with the following mathematical model as

$$\frac{\sigma_0 t}{\sigma_0 - \sigma} = \frac{1}{ab} + \frac{t}{a} \tag{1}$$

The value assessment of the linearized stress relaxation curve is an asymptotic modulus (E_A). Peleg (1980) suggested that use of asymptotic moduli could be applied to monitor and assess the role of internal structural changes of viscoelastic materials, such as gels. An asymptotic modulus (E_A) can be calculated as (Nussinovitch et al., 1990)

$$E_A = \frac{F_0(1 - a)}{A(\varepsilon)\,\varepsilon} \tag{2}$$

This study focused on the evaluation of asymptotic moduli as an index of the viscoelastic properties of HHP Alaska Pollock and Pacific Whiting surimi gels.

MATERIALS AND METHODS

GEL PREPARATION

Eight hundred grams each of Alaska Pollock and Pacific Whiting surimi were thawed at 2°C to 5°C for 12 hours. The defrosted surimi was chopped and blended at high speed for 5 min in a K7700 Food Processor (Regal Ware, Inc.). During blending, water was added to the surimi paste, and the paste was adjusted to ≈78% moisture (wet basis) to facilitate the extrusion. To inhibit protease activity and assist in gel formation, 1% beef plasma and 2.5% sodium chloride were added. After mixing well, the surimi paste was extruded through a caulking gun into a 25-mm diameter plastic tube. Both ends of the plastic tube were fitted with rubber pistons. The extruded surimi paste was maintained at room temperature (≈22°C) for eight hours.

One tube of surimi gel was kept as a control sol without HHP treatment. The other surimi gels were treated in a laboratory scale high-pressure system (Engineered Pressure Systems, Inc.) at five selected pressures (117, 234, 468, and 585 MPa) and three selected processing times (5, 10, and 15 minutes). The temperature was controlled at 20.5°C. The experiment was replicated three times.

STRESS RELAXATION TEST

After pressure treatment, the asymptotic moduli of HHP surimi gels were analyzed using stress relaxation test. The stress relaxation test was performed with the TA.XT2 Texture Analyser (Texture Technologies Corp.). The 25-mm diameter and 10-mm long HHP surimi gels were compressed to 30% deformation at a cross-head speed of 1 mm/s with a 50-mm diameter flat cylindrical probe. The gels were held for 2 min at 30% of their height. The stress relaxation data were analyzed by the normalized stress relaxation method proposed by Peleg (1979 and 1980). The asymptotic moduli of the HHP surimi, a measure of gel solidity, were then determined from the modified version of Peleg's model (Nussinovitch et al., 1990).

STATISTICAL ANALYSIS

The experimental design was a Randomized Complete Block Design (RBD) with a two-way treatment structure. The treatment pressures were 17, 34, 51, 68, and 85 kpsia, and the treatment times were 5, 10, and 15 minutes. The experiments were replicated three times as batch treatments, where one batch was considered one replication (or one block). A batch of treated gel was subjected to a stress relaxation test after the surimi gel pressurization treatment. The asymptotic moduli were analyzed using Statistical Analysis System Release 6.12 (SAS Institute Inc.). An analysis of variance determined whether the level of pressure or processing time influenced the asymptotic moduli. The analyzed values were considered significant when $P < 0.05$. The least significant differential (LSD) also showed the significant difference ($P < 0.05$) for each treatment. In the resulting figures, the pressure treatments with the same letters were found to be not significantly different at $P < 0.05$.

RESULTS AND DISCUSSION

The asymptotic moduli of Alaska Pollock and Pacific Whiting surimi gels calculated from the stress relaxation curves using Peleg's model were significantly increased ($P < 0.05$) by treatment pressure. The asymptotic moduli of both HHP surimi gels treated at higher than 234 MPa were greater than the control gels produced from both species. The time of pressurization significantly increased ($P < 0.05$) the asymptotic moduli of Alaska Pollock surimi gels but did not significantly affect the asymptotic moduli of the Pacific Whiting surimi gels ($P \ell 0.05$).

The HHP Alaska Pollock and Pacific Whiting surimi gels presented equivalent stress relaxation patterns. Figure 8.1 exhibits the general stress relaxation pattern of the HHP Alaska Pollock and Pacific Whiting gels treated at selected pressures. Higher pressures yielded higher stress level curves that corresponded to the asymptotic moduli in the normalized stress relaxation model proposed by Peleg (1980). At the 117 to 585 MPa pressure range, the asymptotic moduli varied from 25.8 to 44.9, 25.9 to 44.7, and 25.7 to 44.6 KPa for Alaska Pollock [Figure 8.2(a)], and 20.9 to 35.7, 20.3 to 36.9, and 22.1 to 36.5 KPa for Pacific Whiting [Figure 8.2(b)], for 5, 10, and 15 minutes, respectively. The asymptotic moduli of both species increased with an increase in pressure. The increase in asymptotic moduli suggests that, in both Alaska Pollock and Pacific Whiting

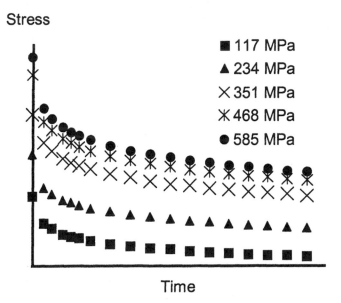

Figure 8.1 General pattern of stress relaxation curves for Alaska Pollock and Pacific Whiting surimi gels treated at selected pressures.

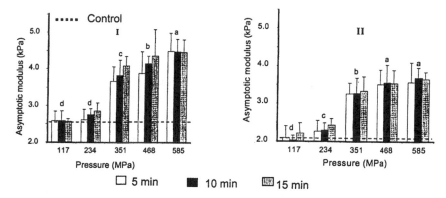

Figure 8.2 Asymptotic moduli of HHP Alaska Pollock (I) and Pacific Whiting (II) surimi gels calculated from Peleg's model. Means among pressure treatments followed by the same character are not significantly different. The error bars represent the standard deviations of the experimental data, and the dotted line refers to the surimi sol without HHP treatment.

surimi, the residual stress at equilibrium (asymptotic) increases when the processing pressure is increased.

The use of Peleg's model was justified for the relaxation curve prediction. Table 8.1 presents the parameters a and b used to fit the normalized stress relaxation curve. In all cases, the stress relaxation curves fit well with Peleg's model ($r^2 \approx 0.99$).

TABLE 8.1. Parameters a and b for the Stress Relaxation Curve Prediction of HHP Surimi Gels by Peleg's Model.

Pressure (Kpsia)	Time (min)	Alaska Pollock			Pacific Whiting		
		a	b	r^2	a	b	r^2
17	5	0.298	3.68E + 06	0.987	0.417	2.99E + 06	0.987
	10	0.315	3.69E + 06	0.985	0.409	2.90E + 06	0.988
	15	0.323	3.66E + 06	0.984	0.420	3.16E + 06	0.985
34	5	0.318	3.73E + 06	0.986	0.423	3.24E + 06	0.988
	10	0.328	3.92E + 06	0.984	0.417	3.28E + 06	0.988
	15	0.321	4.07E + 06	0.986	0.409	3.47E + 06	0.988
51	5	0.337	5.24E + 06	0.988	0.408	4.66E + 06	0.989
	10	0.347	5.46E + 06	0.987	0.402	4.68E + 06	0.989
	15	0.341	5.84E + 06	0.988	0.396	4.77E + 06	0.989
68	5	0.365	5.54E + 06	0.988	0.424	5.02E + 06	0.987
	10	0.359	5.92E + 06	0.988	0.406	5.08E + 06	0.989
	15	0.362	6.22E + 06	0.987	0.404	5.05E + 06	0.988
85	5	0.363	6.41E + 06	0.986	0.428	5.10E + 06	0.988
	10	0.358	6.39E + 06	0.988	0.419	5.27E + 06	0.987
	15	0.361	6.37E + 06	0.988	0.411	5.21E + 06	0.988

The significance of the treatment time of Alaska Pollock surimi gels is useful for optimizing the process design of HHP surimi-based product development. The production of kamaboko (Japanese fish cake) with a desirable asymptotic moduli of about 42 KPa can be accomplished by two methods: first, by using the short time treatment at high pressure; or second, by using long time treatment at low pressure. The appropriate process may be selected depending on the economics of processing and the quality of surimi gels with respect to other desired properties, such as whiteness or cohesiveness.

Asymptotic moduli may be used as an index of HHP surimi gel quality. The results indicate that higher asymptotic moduli exhibit higher resistance to stress at constant deformation. Kamaboko with high asymptotic moduli may thus be more desirable to consumers. The results also demonstrate a potential for the use of HHP in the manufacturing of gels at room temperature. High-pressure treatment at low temperature preserves the nutritional aspects of novel products and retains the fresh-like flavor that is typically encountered in heat-processed gels.

CONCLUSION

Compared to the control surimi sol, the increase in asymptotic moduli of HHP Alaska Pollock and Pacific Whiting surimi gels indicated that the quality of surimi gel improved with an increase in the treatment pressure range used in this study (117 to 585 MPa). As the pressure increased, the asymptotic moduli of both gels significantly increased ($P < 0.05$). The significance of the treatment time in Alaska Pollock surimi gels is useful for HHP surimi-based product development in optimizing process design. Expectations for surimi-based products, such as kamaboko using high-pressure technology, will possess a unique quality, such as high asymptotic moduli at high treatment pressure, and may be used accompanied by the heat treatment to assist gel formation in surimi-based products.

NOMENCLATURE

a = asymptotic level constant
b = initial relaxation rate constant
$A(\epsilon)$ = calculated cross-sectional area of the relaxing gel
F = initial force
L = length
σ = decreasing stress at time t

σ_0 = initial stress
ϵ = imposed strain
ϕ = diameter

REFERENCES

Bertak, J. A. and Karahadian, C. 1995. Surimi-based imitation crab characteristics affected by heating method and end point temperature. *J. Food Sci.* 60 (2):292–296.

Hoover, D. G. 1993. Pressure effects on biological systems. *Food Technol.* 47 (6):150–155.

Kamath, G. G., Lanier, T. C., Foegeding, E. A., and Hamann, D. D. 1992. Nondisulfide covalent cross-linking of myosin heavy chain in "setting" of Alaska Pollock and Atlantic Crocker surimi. *J. Food Biochem.* 16 (2):151–172.

Kamazawa, Y., Numazwa, T., Seguro, K., and Motoki, M. 1995. Suppression of surimi gels setting by transglutaminase inhibitors. *J. Food Sci.* 60 (4):715–717, 726.

Lanier, T. C. 1996. Gelation of surimi pastes treated by high hydrostatic pressure. In *High Pressure Bioscience and Biotechnology: Proceedings of the International Conference on High Pressure Bioscience and Biotechnology,* Kyoto, Japan. November 5–9, 1995. Elsevier Science, Oxford. pp. 357–362.

Mohsenin, N. N. 1986. Rheological properties. In *Physical Properties of Plant and Animal Materials.* Gordon and Breach Science Publishers Inc., Canada.

Nussinovitch, A., Kaletunc, G., Normand, M. D., and Peleg, M. 1990. Recoverable work versus asymptotic relaxation modulus in agar, carrageenan and gellan gels. *J. Texture Studies.* 21 (4):427–438.

Ohmiya, K., Fukami, K., Shimizu, S., and Gekko, K. 1987. Milk curdling by rennet under high pressure. *J. Food Sci.* 52 (1):84–87.

Okamoto, M., Kawamura, Y., and Hayashi, R. 1990. Application of high pressure to food processing: Textural comparison of pressure- and heat-induced gels of food proteins. *Agric. Biol. Chem.* 54 (1):183–189.

Peleg, M. 1979. Characterization of the stress relaxation curves of solid foods. *J. Food Sci.* 44 (1):277–281.

Peleg, M. 1980. Linearization of relaxation of creep curves of solid biological materials. *J. Rheol.* 24 (4):451–463.

Shoji, T., Saeki, H., Wakameda, A., Nakamura, M., and Nonaka, M. 1990. Gelation of salted paste of Alaska Pollock by high hydrostatic pressure and change in myofribillar protein in it. *Nippon Suisan Gakkaishi.* 56 (12):2069–2076.

Tauscher B. 1994. Pasteurization of food by hydrostatic high pressure: Chemical aspects. *Z. Lebensm Unters Forsch.* 200:3–13.

Van Camp, J., and Huyghebaert, A. 1995. High pressure-induced gel formation of a whey protein and haemoglobin protein concentrate. *Leben. Wiss. Technol.* 28 (1):111–117.

Van Camp, J., Feys, G., and Huyghebaert, A. 1996. High pressure-induced gel formation of haemoglobin and whey proteins at elevated temperatures. *Leben. Wiss. Technol.* 29:49–56.

Yoo, B., and Lee, C. M. 1993. Rheological relationships between surimi sol and gel as affected by ingredients. *J. Food Sci.* 58 (4):880–883.

CHAPTER 9

The Influence of Temperature on Functional Properties of High Hydrostatic Pressure Surimi Gels

JATUPHONG VARITH
GUSTAVO V. BARBOSA-CÁNOVAS
BARRY G. SWANSON

INTRODUCTION

HIGH hydrostatic pressure (HHP) is a promising technology that could be used to replace or complement heat-induced surimi gels. The use of HHP changes important textural properties, such as hardness and cohesiveness. Therefore, careful selection of processing time, temperature, and pressure becomes relevant for optimizing surimi-based product acceptability.

Many researchers (e.g., Cheng et al., 1979; Chung et al., 1994; Chang, 1995; Ma, 1996) have reported on the properties of surimi induced by heat and pressure. However, intensive study of the combination of pressure and temperature during surimi processing has not been investigated. It has been shown that the increase in pressure combined with temperature may improve the quality of denatured protein gels. Whey protein concentrate (WPC) gels treated at 400 MPa for 30 min increased in gel strength when the treatment temperature increased from 20°C to 60°C. The gel strength of HHP WPC was comparable with the heat-set gels at 80°C for 30 min. It is suggested that pressure can be applied primarily to induce protein denaturation. Once denaturation is achieved, temperature may assist in the formation of intermolecular interactions between polypeptide side chains of denatured protein molecules (Van Camp et al., 1996).

Protein denaturation by high pressure can be explained as the relationship between free energy (ΔG) and volume (ΔV) changes (Balny and Masson, 1993; Tauscher, 1995). Figure 1.1 presents the pressure and temperature phase diagram of the native denatured protein equilibrium. The denaturation temperature rises initially with increasing pressure. Beyond the maximum transition

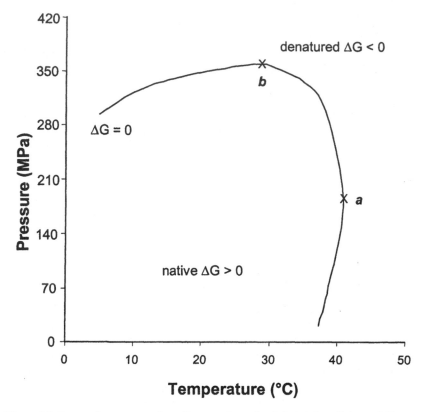

Figure 9.1 Pressure/temperature phase diagram of proteins (adapted from Balny and Masson, 1993). Point (a) refers to the transition temperature and point (b) refers to the transition pressure of protein denaturation.

temperature (point a), the protein denatures at lower temperatures at a given pressure. Beyond the maximum transition pressure (point b), the protein denatures at low pressure at the given temperature. However, the protein in surimi may be denatured at low temperatures and low pressures so that the denatured protein may show special textural properties during testing (Tauscher, 1995).

The objective of this work was to study the effects of temperature on surimi gel functional properties during pressurization, using some well-known physical characterization tests.

MATERIALS AND METHODS

GEL PREPARATION

Eight hundred grams each of Alaska Pollock and Pacific Whiting surimi were thawed at 2°C to 5°C for 12 hours. The defrosted surimi was chopped into 125 cm³ cubes and blended at high speed for 5 min in a K7700 Food Processor (Regal Ware, Inc.). During blending, water was added to the surimi paste, adjusting it to ≈78% moisture (wet basis) to facilitate the extrusion. To inhibit protease activity and assist in gel formation, 1% beef plasma and 2.5% sodium chloride were added. After mixing well, the surimi paste was extruded with a caulking gun into a 25-mm diameter plastic tube where both ends were plugged with rubber pistons. The surimi paste was left at room temperature (≈22°C) for eight hours.

One tube of surimi gel was kept as a control sol without HHP treatment. The other surimi gels were treated by a laboratory scale high-pressure system (Engineered Pressure Systems, Inc.) at five selected temperatures (25, 40, 55, 79, and 85°C) and three selected treatment times (5, 10, and 15 minutes) with a constant pressure of 351 MPa. The experiment was replicated three times. After pressure processing, the functional properties of surimi gels were analyzed by stress relaxation, texture profile analysis, water holding capacity, and whiteness.

STRESS RELAXATION TEST

In the stress relaxation test, the surimi gels were exposed to fixed deformation. The cylindrical surimi gel ($\phi 25$ mm × $L10$ mm) was subjected to stress with the TA.TX2 Texture Analyzer (Texture Technologies Corp.) by moving the 50-mm diameter plunger down at a speed of 1 mm/s and keeping the deformation constant at 30% strain for two minutes. The linearized stress relaxation model proposed by Peleg (1979) was used to simplify the stress relaxation data as follows:

$$\frac{\sigma_0 t}{\sigma_0 - \sigma} = \frac{1}{ab} + \frac{t}{a} \tag{1}$$

The asymptotic modulus of heat-HHP-induced surimi gels was determined by the modified version of Peleg's model (Nussinovitch et al., 1990) as

$$E_A = \frac{F_0(1 - a)}{A(\varepsilon)\,\varepsilon} \tag{2}$$

TEXTURE PROFILE ANALYSIS (TPA)

The TPA was performed with the TA.TX2 Texture Analyzer with a $\phi 50$-mm cylindrical probe. The $\phi 25 \times L10$ mm cylindrical heated HHP surimi gel was compressed twice to 30% of its height at room temperature (20°C to

23°C), and the cross-head speed of compression was set at 1 mm/s. The TPA was replicated three times. The hardness and cohesiveness were calculated from the TPA curves as the peak of the first compression curve and the area under the first compression curve divided by area under the second compression curve, respectively.

WATER HOLDING CAPACITY TEST

Water holding capacity refers to the ability of a food system to retain water in a stable state throughout the shelf-life of a food (Katz, 1997). The heated HHP surimi gel was cut into small pieces, and about 2 g of scrap surimi gel were spread on a ϕ11-cm Whatman #50 filter paper and then backed up with two ϕ11-cm Whatman #1 filter papers. The filter papers were folded twice and centrifuged in a RT6000B Sorvall centrifuge (Dupont Co.) at 2,500 rpm for 15 min. The water holding capacity was calculated with the following equation:

$$WHC\ (\%) = \frac{IWC - WL}{IWC} \times 100 \qquad (3)$$

where IWC can be obtained from the percentage of moisture content (following AOAC, 1991) multiplied by the weight of the gel sample.

COLOR TEST

Color is the easiest property of heated HHP surimi that can be assessed by the naked eye. The change in whiteness of processed surimi confirms that the proteins in the surimi have been denatured. The whiteness of the heat pressure-induced gel sample in this study was measured with the CM2001 Minolta color spectrophotometer. Specifically, L^*, a^*, and b^* of the CIELAB system were used to calculate the whiteness as follows:

$$WN = 100 - [(100 - L^*)^2 + a^{*2} + b^{*2}]^{1/2} \qquad (4)$$

STATISTICAL ANALYSIS

The investigation of temperature and time effects during pressurization was designed as a randomized complete block design (RBD) using a two-way treatment structure (temperature and time). The treatment temperatures were selected as 25, 40, 55, 70, and 85°C, while the treatment times were 5, 10, and 15 min. The experiments were completed in blocks of three replications. One replication refers to one batch of treatment completed by five selected treatment temperatures and three selected treatment times. After each replication, rheological, textural, and miscellaneous tests were performed. The data (asymptotic moduli, hardness, cohesiveness, water holding capacity, and whiteness) were analyzed by the Statistical Analysis System Release 6.12 (SAS Institute, Inc.). An analysis of variance determined whether the level of pressure or processing time influenced the gel properties. The analyzed values were considered significant when $P < 0.05$. The least significant differential (LSD) also showed the significant difference ($P < 0.05$) for each treatment. In the resulting figures, the pressure treatments with the same letters are not significantly different at $P < 0.05$.

RESULTS AND DISCUSSION

ASYMPTOTIC MODULI

The asymptotic moduli of heated HHP Alaska Pollock and Pacific Whiting surimi gels were significantly increased ($P < 0.05$) by treatment temperatures up to 70°C. The treatment time was significant ($P < 0.05$) for the asymptotic moduli of Pacific Whiting surimi gels but not significant ($P \geq 0.05$) for asymptotic moduli of Alaska Pollock surimi gels. All heated HHP surimi gels exhibited greater asymptotic moduli than the control sols. The asymptotic moduli of heat-HHP Alaska Pollock and Pacific Whiting surimi gels are presented in Figures 9.2(a) and 9.2(b).

The asymptotic moduli of Alaska Pollock gels increased from about 29 KPa at 25°C to 47 KPa at 85°C. The increase in asymptotic moduli indicated that the rigidity of Alaska Pollock gels increased due to the treatment temperature during pressurization. For Pacific Whiting gels, the asymptotic moduli had a transition temperature (the point at which the moduli trend changed from increasing to decreasing) that depended on time. The asymptotic moduli of Pacific Whiting gels increased as the temperature increased until it reached the transition temperature of 70°C for the gels treated for 5 min and 55°C for those treated for 10 and 15 min (Figure 9.2). Beyond the transition temperature, the asymptotic moduli of the Pacific Whiting surimi gels decreased. The moduli for gels treated for long times (10 to 15 min) at high temperature (> 55°C) significantly decreased ($P < 0.05$). The decrease in asymptotic moduli of Pacific Whiting surimi gels treated at higher than 55°C suggests that Pacific Whiting should not be processed for a long time at high temperatures during pressurization. The possible reason for this decrease is the activation of proteolytic enzymes in Pacific Whiting surimi. Tauschei (1995) stated that the combination of temperature and pressure shifts the chemical reaction equilibrium (in this case, enzymatic reaction equilibrium), so a shift in the enzymatic reaction equilibrium due to the combination of temperature and pressure may

TABLE 9.1. Parameters a and b for Stress Relaxation Curve Prediction of HHP Surimi Gels by Peleg's Model [Equation (1)].

Temperature (°C)	Time (min)	Alaska Pollock			Pacific Whiting		
		a	b	r^2	a	b	r^2
25	5	0.384	0.028	0.984	0.494	0.037	0.985
	10	0.406	0.028	0.986	0.499	0.034	0.986
	15	0.427	0.030	0.984	0.489	0.037	0.987
40	5	0.423	0.031	0.982	0.495	0.035	0.986
	10	0.414	0.028	0.986	0.500	0.034	0.986
	15	0.405	0.028	0.987	0.498	0.037	0.985
55	5	0.403	0.029	0.984	0.452	0.035	0.985
	10	0.410	0.029	0.985	0.463	0.035	0.984
	15	0.406	0.030	0.985	0.473	0.035	0.984
70	5	0.402	0.030	0.984	0.444	0.036	0.984
	10	0.414	0.033	0.984	0.478	0.041	0.981
	15	0.399	0.031	0.986	0.482	0.042	0.982
85	5	0.400	0.033	0.985	0.468	0.041	0.983
	10	0.421	0.036	0.983	0.554	0.053	0.981
	15	0.395	0.033	0.986	0.627	0.065	0.981

activate the proteolytic enzyme in Pacific Whiting gels. There are many published papers on the typical behavior of enzymes under HHP at different temperatures (Ko et al., 1991; Knorr, 1993); however, there is no recent published paper supporting enzymatic activity in proteolytic enzymes for heated HHP Pacific Whiting surimi gels.

Regarding the prediction of stress relaxation curves, the use of a linearized approach by Peleg (1979) was justified. The predicted relaxation curves fitted well with average $r^2 = 0.98$. Table 9.1 presents the parameters a and b used in the linearized model [Equation (1)] and r^2 for each curve prediction.

HARDNESS AND COHESIVENESS

The hardness of Alaska Pollock and Pacific Whiting surimi during 351 MPa pressurization was significantly affected ($P < 0.05$) by treatment temperature. The treatment time had a significant effect ($P < 0.05$) on the hardness of Pacific Whiting gels, but not on that of Alaska Pollock gels ($P \geq 0.05$) (see Figure 7.3).

The average hardness of heated HHP Alaska Pollock significantly increased ($P < 0.05$) from 11.80 N at 25°C to 18.91 N at 85°C. For Pacific Whiting gel, the hardness significantly increased ($P < 0.05$) from 11.14 N to 16.27 N at temperatures lower than 55°C. At temperatures higher than 55°C, the hardness of Pacific Whiting gels treated for 5 min still increased, but the hardness of gels treated for 10 or 15 min decreased. At 85°C, the hardness of Pacific

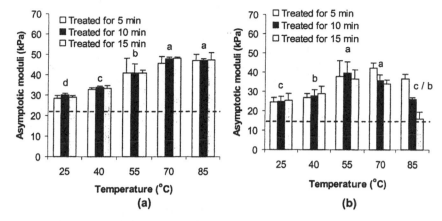

Figure 9.2 Asymptotic moduli of Alaska Pollock (a) and Pacific Whiting (b) surimi gels treated at selected temperatures under 351 MPa pressure. Means among temperature treatments followed by the same character are not significantly different. The error bars represent the standard deviations of the experimental data, and the doted line refers to the surimi sol without HHP treatment.

Whiting gels was 16.31, 14.14, and 10.76 N for treatments of 5, 10, and 15 min, respectively. These results indicate that at temperatures greater than 55°C, the hardness of Pacific Whiting gels significantly decreased ($P < 0.05$), depending on treatment time.

The cohesiveness, i.e., the strength of internal bonds making up the body of a material, was significantly decreased ($P < 0.05$) by treatment temperatures for both Alaska Pollock and Pacific Whiting surimi gels. The treatment time significantly affected ($P < 0.05$) the cohesiveness of heated HHP Pacific Whiting gels, but did not significantly affect ($P \geq 0.05$) the cohesiveness of heated HHP Alaska Pollock gels (Figure 9.4).

The average cohesiveness of Alaska Pollock gels at 25°C was 0.89. As the temperature was increased to 85°C, the cohesiveness of Alaska Pollock gels decreased to 0.86, regardless of time. The effects of treatment temperature and time on the cohesiveness were more obvious in Pacific Whiting gels than in Alaska Pollock gels. When the temperature of treatment increased from 25°C to 55°C, the cohesiveness of Pacific Whiting gels did not significantly change ($P \geq 0.05$). At treatment temperatures higher than 55°C, the cohesiveness of Pacific Whiting gels significantly decreased ($P < 0.05$) depending on time. At 85°C, the Pacific Whiting gels treated for 15 min exhibited the smallest cohesiveness, which was attributed to loose bonding within the proteins in gels. The loose bonding of Pacific Whiting gels treated at 85°C were soft (no strength or hardness) to the touch and collapsed easily when low force was applied with a finger.

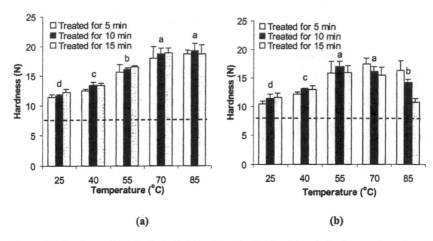

Figure 9.3 Hardness of Alaska Pollock (a) and Pacific Whiting (b) surimi gels treated at selected temperatures under 351 MPa pressure. Means among temperature treatments followed by the same character are not significantly different. The error bars represent the standard deviations of the experimental data and the doted line refers to the surimi sol without HHP treatment.

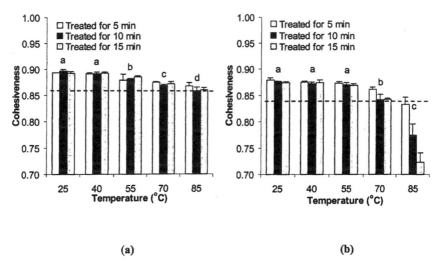

Figure 9.4 Cohesiveness of Alaska Pollock (a) and Pacific Whiting (b) surimi gels treated at selected temperatures under 351 MPa pressure. Means among temperature treatments followed by the same character are not significantly different. The error bars represent the standard deviations of the experimental data and the doted line refers to the surimi sol without HHP treatment.

118

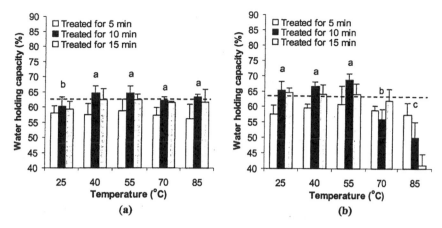

Figure 9.5 Water holding capacity of Alaska Pollock (a) and Pacific Whiting (b) surimi gels treated at selected temperatures under 351 MPa pressure. Means among temperature treatments followed by the same character are not significantly different. The error bars represent the standard deviations of the experimental data, and the doted line refers to the surimi sol without HHP treatment.

The decrease in hardness and cohesiveness of heated HHP Pacific Whiting gels at treatment temperatures higher than 55°C may be related to their proteolytic activity, which is more severe at high temperatures (85°C) than at intermediate temperatures. Because the acceptable range of cohesiveness for surimi gels based on our experience was 0.80 to 0.95, the heated HHP Pacific Whiting gels treated at 85°C for 10 and 15 min (cohesiveness of 0.77 and 0.72, respectively) were weak and not acceptable. However, additional sensory tests are needed to confirm the acceptability of some heated HHP surimi gels.

WATER HOLDING CAPACITY

The water holding capacity of heated HHP-induced Alaska Pollock gels was not significantly affected ($P \geq 0.05$) by treatment temperatures, but was significantly affected ($P < 0.05$) by treatment times. Figure 9.5 shows that heated HHP-induced Alaska Pollock gels treated for 10 min exhibited the highest water holding capacity (62.9%), followed by those treated for 15 and 5 min (61.4% and 57.5%, respectively). This implies that heated HHP Alaska Pollock surimi gels containing fine protein networks bind the water consistently with respect to selected treatment temperatures.

For the Pacific Whiting gels, both temperature and time during pressurization exhibited a significant effect ($P < 0.05$) on water holding capacity. With the elevation in temperature from 25°C to 55°C, the water holding capacity of gels treated for 5 and 10 min consistently increased. At temperatures higher than 55°C, the water holding capacity of Pacific Whiting gels decreased. The

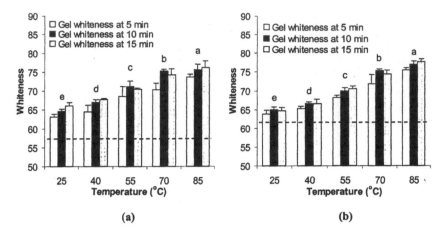

Figure 9.6 Whiteness of Alaska Pollock (a) and Pacific Whiting (b) surimi gels treated at selected temperatures under 351 MPa pressure. Means among temperature treatments followed by the same character are not significantly different. The error bars represent the standard deviations of the experimental data, and the doted line refers to the surimi sol without HHP treatment.

gels treated for 5 min exhibited the lowest degree of reduction among the 5-, 10-, or 15-minute treatment times.

The most appropriate temperature range for heated HHP-induced Alaska Pollock and Pacific Whiting surimi gels that yielded the highest water holding capacity was 40°C–55°C for 10 min. At high treatment temperatures, such as 70°C to 85°C, the water holding capacity of Pacific Whiting gels declined, producing poor water holding capacity and weak gel networks, corresponding to a decrease in other properties, such as strength and cohesiveness. Based on our experience, a water holding capacity less than 50% is not acceptable, because it may shorten shelf-stability. Therefore, Pacific Whiting surimi gels treated for 10 and 15 min at 85°C during a 351 MPa HHP treatment (with 50% and 41% water holding capacity) are not acceptable. Additional shelf-stable tests are needed to confirm the acceptability of heated HHP Pacific Whiting surimi gels as determined by their water holding capacity.

WHITENESS

The heated HHP surimi gels produced from both species were transparent at temperatures below 25°C and became more opaque when temperatures were raised to 85°C. The whiteness of heated HHP Alaska Pollock and Pacific Whiting gels was significantly increased ($P < 0.05$) by both treatment temperature and treatment time during pressurization at 351 MPa (Figure 9.6). The change in whiteness of the heated HHP surimi gels indicates that the fish muscle protein may have been denatured. The whiteness of both surimi gels treated for 5 min was less than the whiteness of surimi gels treated for 10 or 15 min ($P < 0.05$). However, there was no significant ($P \geq 0.05$) difference between the whiteness of gels treated for 5 or 10 minutes. The whiteness of Alaska Pollock was between 63.1% and 76.1%, while the whiteness of Pacific Whiting varied from 63.7% to 77.7%. Because the heated HHP surimi gels exhibited a maximum whiteness at about 77%, the heated HHP surimi gel may need additional formulation to reach the expected 78% whiteness of commercial kamaboko.

CONCLUSIONS

When HHP is combined with heat treatment, the functional properties of induced Alaska Pollock surimi gels are improved. Similar to conventional heated-induced surimi gels, the functional properties of heated HHP Alaska Pollock were significantly improved ($P < 0.05$) with increases in temperatures, except for the water holding capacity, which decreased as the temperatures increased. For Pacific Whiting surimi gels, the functional properties were improved as the temperature increased to a transition temperature of 55°C under 351 MPa pressurization. At temperatures higher than 55°C, the functional properties of Pacific Whiting surimi gels deteriorated, except for whiteness, which increased as the temperatures increased. One possible reason for the deterioration of the functional properties of HHP Pacific Whiting gels is that protease activity may occur. Heat combined with HHP may activate the enzymatic reaction at temperatures higher than 55°C and may lead to deterioration of the functional properties of Pacific Whiting surimi gels, which is more severe at a high temperature, such as 85°C, than at a low temperature of 55°C. The deterioration of heated HHP Pacific Whiting surimi gels suggests that the combination of heat and HHP at temperatures greater than 55°C had a negative impact on the quality of Pacific Whiting surimi gels. The treatment time at temperatures greater than 55°C also affected the functional properties of Pacific Whiting surimi gels. The longer the treatment time at 85°C, the poorer the quality of heated HHP Pacific Whiting surimi gels.

NOMENCLATURE

a = asymptotic level constant
b = initial relaxation rate constant
a^* = scale of redness (+) to greenness (−)
b^* = scale of yellowness (+) to blueness (−)
$A(\epsilon)$ = corresponding calculated cross-sectional area of the relaxing gel
L = length
L^* = lightness on a 0 to 100 scale from black to white

IWC = initial water content
WHC = water holding capacity
WL = water loss
Wn = whiteness
σ = decreasing stress at time t
σ_0 = initial stress
ϵ = imposed strain
ϕ = diameter

REFERENCES

AOAC. 1991. Official Methods of Analysis of the Association of Official Analytical Chemists. Washington, DC.

Balny, C., and Masson, P. 1993. Effects of high pressure on proteins. *Food Reviews International.* 9 (4):611–628.

Chang, S. 1995. Textural properties of Pacific Whiting surimi gels. M.S. thesis, Washington State University, Pullman, WA.

Cheng, C. S., Hamann, D. D., and Webb, N. B. 1979. Effect of thermal processing on minced fished gel texture. *J. Food Sci.* 44 (4):1080–1086.

Chung, Y. C., Gebrehiwot, A., Farkas, D. F., and Morrissey, M. T. 1994. Gelation of surimi by high hydrostatic pressure. *J. Food Sci.* 59 (3):523–524, 543.

Katz, F. 1997. The changing role of water binding. *Food Technol.* 51 (10):64–66.

Knorr, D. 1993. Effects of high hydrostatic pressure processes on food safety and quality. *Food Technol.* 47 (6):156, 158–161.

Ko, W. C., Tanaka, M., Nagashima, Y., Taguchi, T., and Amano, K. 1991. Effect of pressure treatment of actomyosin. ATPases from flying fish and sardine muscles. *J. Food Sci.* 56 (2):338–340.

Ma, L. 1996. Viscoelastic characterization of surimi gel: Effect of setting and starch. *J. Food Sci.* 61 (5):881–883.

Nussinovitch, A., Ak, M. M., Normand, M. D., and Peleg, M. 1990. Characterization of gellan gels by uniaxial compression, stress relaxation and creep. *J. Texture Studies.* 21 (1):37–49.

Peleg, M. 1979. Characterization of the stress relaxation curves of solid foods. *J. Food Sci.* 44 (1):277–281.

Tauscher, B. 1995. Pasteurization of food by hydrostatic high pressure: Chemical aspects. *Z. Lebensm Unters Forsch.* 200:3–13.

Van Camp, J., Feys, G., and Huyghebaert, A. 1996. High pressure-induced gel formation of haemoglobin and whey proteins at elevated temperatures. *Leben. Wiss. Technol.* 29:49–56.

CHAPTER 10

Minimally Processed Fruits Using Hurdle Technology

JORGE WELTI-CHANES
STELLA M. ALZAMORA
AURELIO LÓPEZ-MALO
MARIA S. TAPIA

THE consumption of minimally processed foods (MPF) is noticeably increasing in the world market and strongly related to these types of products are those referred to as "partially processed foods (PPF)." Within this group, high moisture fruits seem to be a good choice for the commercialization of tropical and subtropical fruits from around the world.

MINIMALLY PROCESSED FOODS

Minimally processed foods and partially processed foods (fresh-like foods) began to appear on the market at the industrial level in the late 1980s and early 1990s as an answer to the needs of consumers who were interested in products that were fresh, as well as easy to prepare and serve (Resurrección and Prussia, 1986; Mertens and Knorr, 1992). The appearance of these types of products has been closely related to both changes in consumption patterns of society (Tapia de Daza et al., 1996) and certain needs of the catering industry (Ahvenainen, 1996). In many countries where there are no storage or transport refrigeration facilities, MPF and PPF may act as a mechanism to regulate fruit and vegetable production and their supply to final transformation industries (Alzamora et al., 1993; Alzamora et al., 1995; Argaiz et al., 1995).

From an economic point of view, the growth of MPF and PPF consumption in European markets has been gradually increasing over the past few years (Dave and Gorns, 1993), while, for the United States, it is foreseen that 25% of the food market will be for these types of products by the year 2000. In

Latin America, the number of PPF and MPF products appearing on the market at reasonable prices for consumers has steadily increased.

MPF and PPF include products and processes that may be grouped in diverse categories such as MPF with invisible processing, carefully processed, partially processed, and high moisture shelf-stable. All these terms represent the evolution of definitions that have been given since the original concepts set forth by Rolle and Chism (1987), the modifications made by Shewfelt (1987), Huxsoll and Bollin (1989), Wiley (1994), and Ohlsson (1994), and the new concepts presented by Tapia de Daza et al. (1996) for minimally processed fruits based on the combination of preservation factors or hurdle technology proposed by Leistner (1985, 1992) and Gould (1989).

An early definition of minimal processing considered only those products that maintain their freshness by keeping the biological tissues alive (Rolle and Chism, 1987; Shewfelt, 1987), but the current definition considers those products that maintain the characteristics of fresh foods (or close to them) by inactivating the cellular metabolism in biological tissues (Huxsoll and Bollin, 1989; Wiley, 1994; Ohlson, 1994). Taking into account this latest version, high-moisture fruit products (HMFP) preserved by hurdle technology can be classified as MPF.

The definition of HMFP is related to that given by Wiley (1994) for minimally processed refrigerated fruits (MPRF). According to this author, the differences between MPRF and fresh fruits or those that have been preserved by cold, irradiation, dehydration, or thermal treatments are based on four main aspects: (1) quality of the product; (2) preservation method; (3) storage form and conditions; and (4) packaging procedure. MPRF keep their freshness characteristics to a greater extent than any of the other fruit products mentioned by Wiley, having an expected average shelf-life of around 21 days or greater, compared to that of an MPF (4 to 7 days). Both types of products should comply with the key requirements mentioned by Ahvenainen (1996) to obtain high-quality MPF. The only difference between MPF and MPRF is the use of refrigeration as a preservation factor.

In addition to the preparation and manipulation steps of MPF, HMFP make use of other preservation factors, such as blanching, reduction of water activity (a_w) and pH, and incorporation of antimicrobial agents and other additives that substitute for the need for refrigeration. HMFP represent an improved version of many intermediate moisture products and a variant of the MPRF in which refrigeration is not necessary to increase the stability and shelf life of the product. Table 10.1 shows a comparison among three types of products developed from fruits (Tapia de Daza et al., 1996). It can be observed that the a_w levels of MPRF and HMFP are similar and, in general, higher than those of classic intermediate moisture fruits. In HMFP, the incorporation of a_w depressing solutes should not affect the freshness of the fruit, and not having to use refrigeration as a preservation factor is seen as an economic

TABLE 1C.1. Comparison of Three Fruit Preservation Systems with Reference to Some Process Characteristics of Products and Processes.

Fruit Process/ Technology	a_w	Overall Quality	Shelf Stability	Preservatives Added	Processes and Preservation Operations	Blanching	Packaging
Intermediate Moisture Fruits (IMF)	0.75–0.92	Slightly modified to modified	Usually shelf life at room temperature	Sulfites, sorbic, benzoic, and ascorbic acid	Peeling, coring, slicing, dipping in preservative solutions, dehydration	Generally required	May be required
Minimally Processed Refrigerated Fruits (MPRF)	0.97–0.99	Fresh-like	Refrigeration required	Might include some (i.e., ascorbic acid)	Peeling, coring, slicing	Possibly, though excluded in most descriptions	Required (modified or controlled atmosphere packaging may be used)
Minimally Processed High Moisture Fruit Products (HMFP)	0.93–0.98	Fresh-like to slightly modified	Shelf life at room temperature	Sulfites; sorbic, citric, benzoic, and ascorbic acid	Peeling, coring, slicing, dipping in preservative solutions	Generally applied	Required

125

and technological objective. Also, the use of other preservation factors (additive incorporation and blanching) distinguishes HMFP from MPRF. In addition, is important to observe that the former does not require special packaging systems.

STABILITY PROBLEMS PRESENTED BY MINIMALLY PROCESSED FOODS WITH SPECIAL EMPHASIS ON FRUITS AND VEGETABLES

When generating a new type of product, as in the case of HMFP, quality changes and possible areas where loss of stability may occur should be clearly recognized. Ahvenainen (1996) stated that the quality modifications of MPF vegetable products may be focused on (1) physiological and biochemical changes; (2) microbiological changes; (3) safety aspects; and (4) nutritional changes. In the case of HMFP, due to the employment of such preservation factors as a_w, pH, blanching, and preservative incorporation, it is necessary to include sensory changes and possible textural and structural modifications.

Three research projects related to the development of adequate technologies for HMFP production were carried out as an attempt to solve the problems linked to the points mentioned above. The first of these projects, "Bulk Preservation of Fruits by Combined Methods Technology," was performed from 1991 to 1994 with eight Ibero-American participant countries (Welti-Chanes and Vergara, 1995). The other two projects that are ongoing began in 1995 with the participation of five Ibero-American countries. These projects are "Development of Minimal Processing Technologies for Food Preservation" and "Development of Minimally Processed Products from Tropical Fruits Using Vacuum Impregnation Techniques."

The development of a simple technology that has been achieved through these projects and its flowchart is presented in Figure 10.1. It can be noticed that the sequence and number of operations are very simple and that the preservation factors (hurdles) employed are those mentioned above (a_w, pH, incorporation of additives, and mild blanching). The flow-chart shows the case of fruit pieces, but when the fruit is transformed into a purée, a pulping stage is included before adjusting the conditions of the final product. The types of hurdles, as well as their levels, are fixed within the ranges shown in Figure 10.1 (Alzamora et al., 1993). Water activity reduction has been carried out mainly by osmotic dehydration in sucrose syrups. At the present time, vacuum osmotic dehydration (Fito and Chiralt, 1995, 1997) is also being applied to reduce process time and improve the quality of the products (Consuegra et al., 1997).

The fruit products developed in the different countries in pieces or in

MANGO, PAPAYA, PINEAPPLE, PEACH

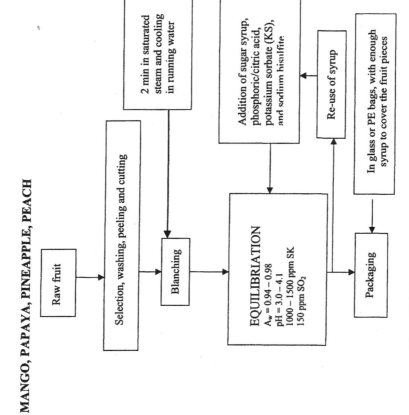

Figure 10.1 Flow-chart to produce shelf-stable high moisture fruits.

127

purée form during the first-mentioned project were apple, banana, citric fruits, chirimoya, fig, guava, kiwi, mango, papaya, passion fruit, peach, pineapple, sapodilla, soursop, strawberry, and tamarind. Many of these products have been improved in an attempt to move closer to the MPF concept by using new preservation factors, reducing antimicrobial and other additives levels, and, in some cases, incorporating the use of refrigeration as an additional factor.

The results obtained in the mentioned projects to obtain HMFP in the form of purées or pieces are presented next and are linked to their quality modification during processing and storage.

MICROBIOLOGICAL CHANGES AND SAFETY ASPECTS

In HMFP, pH plays a more important role in final stability compared to pasteurized fruits and vegetables (Tapia de Daza et al., 1995) or to those products named in a generic way by Leistner (1985, 1992) as pH-SSP. The a_w values of HMFP are in general higher than 0.95, so the pH value, the blanching treatment, and the addition of antimicrobial agents are the set of hurdles that help to avoid microbiological problems. The deteriorative microorganisms that may cause more problems to HMFP are the osmophilic preservative-resistant yeasts, *Zygosaccharomyces rouxii, Z. bailii*, and *Schizosaccharomyces pombe* (Tapia de Daza et al., 1996). Problems related to pathogenic microorganisms are not likely to occur in these types of products, mainly due to the low pH values (3.0 to 4.0) and the types of packages employed, ones that do not promote the anaerobic conditions that can encourage toxigenic microorganisms to grow.

The microbial stability of HMFP has been evaluated by (1) evaluation of the physiological response of selected microorganisms to several stress factors; (2) evolution of the native flora in the developed products; and (3) microbial challenge tests in the products of interest.

Some microbial challenge tests were performed in laboratory model systems with similar HMFP characteristics. The effects of a_w levels (0.95 to 0.99) adjusted with diverse types of sugars, pH (3 to 5), potassium sorbate (0 to 1000 ppm), and, in some cases, sulfites have been evaluated by Mazzota et al. (1990), Cerruti et al. (1988, 1990), Tapia de Daza et al. (1996), and López-Malo et al. (1992). The results of such studies are presented in Table 10.2. It can be observed that in some cases the combination of pH and a_w is enough to inhibit the growth of microorganisms such as *Bacillus coagulans*, but, in other cases, the incorporation of an antimicrobial agent is necessary to avoid the growth of microorganisms such as *S. cerevisiae, A. flavus,* and *A. ochraceus. Z. bailii* is a clear example of a yeast that requires a significant decrease in the a_w level (lower than 0.95) to prevent its growth. At this a_w level, a fruit product

TABLE 10.2. Some Inhibitory Combinations of Preservation Factors in Laboratory Media of *Bacillus coagulans* and Important Fungi in HMFP.

Organism	Inhibitory Combination of Preservation Factors
Bacillus coagulans ATCC 8038	a_w 0.98, pH 4.4
	a_w 0.97, pH 4.9
Saccharomyces cerevisiae	a_w 0.97, pH 5.0, 1,000 ppm SK
	a_w 0.97, pH 4.0, 500 ppm SK
	a_w 0.97, pH 3.5, 200 ppm SK
	a_w 0.95, pH 3.01, 50 ppm SK, 50 ppm SO_2, 50–55°C
Zygosaccharomyces rouxii	a_w 0.95, pH 4.0, 500 ppm SO_2
Aspergillus flavus	a_w 0.99–0.97, pH 4.5, 500 ppm SK,
	a_w 0.95, pH 3.5, 500 ppm SK
Penicillium sp.	a_w 0.99–0.95, pH 4.5, 500 ppm SK
Aspergillus ochraceus SK	a_w 0.99–0.95, pH 4.5, 1,000 ppm

does not keep its freshness and cannot be classified as MPF (Tapia de Daza et al., 1995).

It has been demonstrated in native flora evolution studies of HMFP that the developed products are stable at 25°C and 35°C for periods up to 4 months (Tapia de Daza et al., 1995, 1996); however, possible contamination of these products with osmo-resistant/osmophilic yeasts that may develop preservative resistance may be an important problem. This was demonstrated by Díaz de Tablante et al. (1993) and Corte et al. (1994) by challenging mango (0.97 a_w, pH 3.35, 150 ppm sodium bisulfite, and 1,000 ppm potassium sorbate) and papaya (0.98 a_w, pH 3.5, 150 ppm sodium bisulfite, and 1,000 ppm potassium sorbate) with *Z. bailii*. The results obtained for these products showed that *Z. bailii* might survive and grow under these conditions.

These findings promoted investigation of new hurdles to avoid microbial growth and ensure the freshness of fruits. Palou et al. (1997) specifically studied the application of high hydrostatic pressure (HHP) treatments to inactivate microorganisms such as *Z. bailii* in laboratory model systems with similar characteristics to those used in HMFP. Figure 10.2 shows the results obtained from this experiment when high pressures were applied to model systems (pH 3.5, a_w 0.98, and 1,000 ppm) with or without potassium sorbate in which *Z. bailii* was previously inoculated. It can be observed that the yeast was inactivated with treatments of 4 min or longer at pressures ≥ 517 MPa in the absence of potassium sorbate, but when 1,000 ppm potassium sorbate were added, a shorter process of 2 min at pressures ≥ 345 MPa was employed to yield stable samples in which no cell recovery was observed after the longest incubation periods. Because this kind of treatments theoretically may maintain the flavor

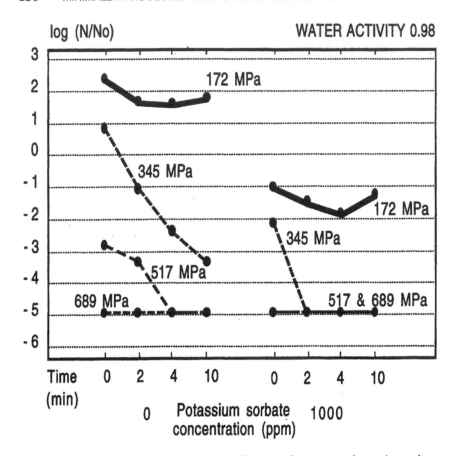

Figure 10.2 Combined effects of high pressure (MPa), time of exposure, and potassium sorbate concentration on the log of *Z. bailii* survival fraction (N/No) at a_w 0.98 and pH 3.5.

and aroma of fruit products, the use of high pressures would be functional when HMFP are developed as pastes or purées.

The need to perform studies on the use of new hurdles and variants of proven methods is evident. For example, it is necessary to amplify the research on the application of nonthermal processes (HHP, electric pulses, ultrasound, etc.) or other antimicrobial agents (those coming from natural sources), and even the reduction of temperature during processing and storage of the product to ensure that growth of polluting microorganisms is avoided.

Research on natural antimicrobials has yielded important results when proving the use of vanillin in model systems similar to HMFP. Table 10.3 shows the important inhibiting effect of vanillin on the growth of various yeasts using concentrations of 1,000 and 2,000 ppm in systems with a_w 0.95 and 0.99 and pH 4.0. This level of vanillin is below the sensory rejection level,

TABLE 10.3. Effect of Vanillin on Growth of *S. cerevisiae, Z. rouxii, Z. bailii,* and *D. hansenii* at pH 4.0 and 27°C.

Yeast a_w	Control		1,000 ppm Vanillin		2,000 ppm Vanillin	
	0.99	0.95	0.99	0.95	0.99	0.95
S. cerevisiae	+	+	+	−	−	−
Z. rouxii	+	+	+	+	−	−
Z. bailii NRRL Y-1448	+	+	+	+	−	−
D. hansenii	+	+	+	−	−	−

+ = growth (increase of CFU exceeding 1 log cycle.)
− = growth (increase of CFU below 1 log cycle).

which is superior to 3,000 ppm (Cerruti and Alzamora, 1996). López-Malo et al. (1997) studied the effect of vanillin on the inhibition of various molds. It was observed that the more resistant mold to vanillin is *Aspergillus niger*, but even in this case, its growth can be inhibited up to 30 days when the system with an a_w 0.98 has 1,000 ppm of vanillin and a pH equal to or smaller than 3.0 and is stored at temperatures equal to or lower than 15°C. These results seem to point out the need to manage temperatures lower than ambient; however, the other molds investigated can be inhibited with vanillin levels around 1,000 ppm at temperatures near 25°C. Similar results have been obtained for systems containing fruit (strawberry, apple, banana, mango, papaya, and pineapple) that may be catalogued as HMFP employing vanillin to inhibit the growth of diverse microorganisms (Cerruti et al., 1997; López-Malo et al., 1995).

The evolution of microbial flora during the elaboration stages of HMFP with pineapple, papaya, mango, and fig was investigated by Alzamora et al. (1989), López-Malo et al. (1994), Díaz de Tablante et al. (1993), and Corona (1991). It was shown respectively that these fruits are, in general, microbiologically stable products. Table 10.4 provides the composition of some HMFP and the storage temperature at which they are microbiologically stable for 3 to 8 months.

Based on the information mentioned above, Tapia de Daza et al. (1996) have proposed a 13-step guide for designing preservation systems to obtain stable and microbiologically safe HMFP using hurdle technology. This guide establishes in simple form the necessity for achieving the quality and stability objectives set for HMFP.

COLOR CHANGES

In the majority of minimally processed (MP) fruit and vegetable products, one of the main deterioration problems is related to enzyme action, which

TABLE 10.4. Shelf-Stable HMF by Hurdle Technology.

Fruit	Hurdles	Storage Temperature (°C)	Shelf Life (Months)
Peach (halves)	a_w = 0.97 (sucrose), pH = 3.7, KS* = 1,000 ppm, NaHSO$_3$ = 150 ppm	35 25	3 8**
Mango (slices)	a_w = 0.97 (sucrose), pH = 3.0, SB† = 1,000 ppm, NaHSO$_3$ = 150 ppm	35	4.5
Papaya (slices)	a_w = 0.97 (sucrose), pH = 3.7, KS = 1,000 ppm	35 25	3 8**
Pineapple (halves)	a_w = 0.97 (sucrose), pH = 3.8, KS = 1,000 ppm, NaHSO$_3$ = 150 ppm	35 25	3 8**
Pineapple (halves or complete)	a_w = 0.97 (glucose), pH = 3.1, KS = 1,000 ppm, NaHSO$_3$ = 150 ppm	27	4**

*Potassium sorbate.
**Shelf life measured only during this period of time, but could be longer.
†Sodium benzoate.

maintains their normal activity. In many cases, it can be increased due to the previous manipulation, especially during cutting. Enzymatic activity mainly affects color and texture due to the food's physiological activity. The blanching stage included in the manufacturing process of HMFP, besides having superficial disinfecting effects on the product, has as its principal objective the reduction of enzymatic activity. However, types of treatments should be very carefully applied because excessive time or exposition temperature can provoke undesirable flavor, aroma, and textural changes in the product that move it away from the MP concept.

The most important enzyme in minimally processed (MP) fruits and vegetables is polyphenol oxidase, which causes browning. The other two enzymes of interest in these products are the pectolytics that cause fruit softness and the lipoxidases that cause oxidation problems with the development of off-flavors. The control of enzymatic activity can be achieved not only by thermal inactivation, but also by making totally or partially unavailable the substrate on which the enzyme acts or, in other cases, avoiding the presence of oxygen or some co-factor needed for the reaction. This is why some schemes that favor the reduction of enzymatic activity by incorporating some chemical compounds or by limiting the quantity of oxygen within the package system of the final product have been used in the development of HMFP.

Color changes linked to the browning of MP products are normally due to the joined action of enzymatic and non-enzymatic reactions that, in many cases, are difficult to distinguish, so the global effect of the studied variables on browning process is normally evaluated. Guerrero et al. (1996) evaluated color changes in banana purée (a_w 0.97, pH 3.1, 1,000 ppm potassium sorbate, 200 ppm ascorbic acid, addition of $NaHSO_3$, and thermal treatment after packaging) during four months of storage as a function of sodium bisulfite concentration (200, 400, 600 ppm), thermal treatment time of the optimum levels of the three variables (400 to 440 ppm of sodium bisulfite, storage temperature from 19 to 36°C, and 1 min thermal treatment) that ensure minimal browning development in the product over the course of four months of storage. The resultant sulfite levels are of particular interest because the use of such compounds has been restricted in much legislation around the world. Therefore, sulfite substitutes to inhibit enzymatic and non-enzymatic browning should be more deeply investigated. For example, in a study performed to control browning in non-blanched apples (a_w 0.96–0.97, pH 3.9–4.2, and 1,000 ppm ascorbic acid), it was found that the color of the fruit remained almost without change during storage at 25°C for 75 days when apples were immersed in 100 ppm of sodium bisulfite solutions and solutions containing 200 ppm of 4-hexylresorcinol (Monsalve-González et al., 1993).

Enzymes that cause color and flavor changes in avocado purée preserved as HMFP can be inactivated using thermal treatments. However, avocado is a highly sensitive product to the development of undesirable flavors when thermally treated. For this reason, some studies have shown that the use of certain chemical compounds could help to avoid browning reactions in these types of products (Sánchez-Pardo et al., 1991; Dorantes-Alvarez, 1997). Similarly, our own preliminary study demonstrated that undesirable color changes can be avoided and polyphenoloxidase in avocado purée (a_w 0.98) partially inactivated to about 15% to 20% of its original activity when treated with high hydrostatic pressure (689 MPa) at three pH levels (3.9, 4.1, and 4.3) (Figure 10.3). This helps to further the case for high pressure as a good option as an additional barrier to be employed in HMFP.

TEXTURAL, STRUCTURAL, AND SENSORY CHANGES

Enzymes that cause textural changes can be inactivated by blanching; however, this factor or hurdle usually has negative effects on the fruit structure even greater than those caused by the action of enzymatic systems (Vidales et al., 1994). The blanching conditions recommended for these purposes should always be carefully applied, and the use of some other hurdle is suggested to inhibit enzyme action. For example, the combined use of mild blanching (30 s with saturated steam) and refrigeration to inhibit the action of pectolytic enzymes in papaya is recommended to improve texture retention in this HMFP (a_w 0.963, pH 3.5).

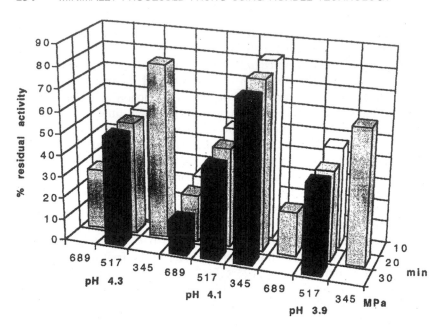

Figure 10.3 Effect of high hydrostatic pressure treatments and initial pH on the residual polyphenol oxidase activity in avocado purée.

The microstructure of HMFP tissue is mainly affected by the blanching treatment and sugar impregnation process to reduce a_w (Alzamora et al., 1997; Vidales et al., 1994). For example, a microscopic evaluation of parenchymatous tissue changes in a strawberry (Vidales et al., 1994) showed that blanching treatments with water as well as steam provokes rupture and complete ultrastructural disorganization of the cellular walls and degradation of the middle lamella. The subsequent impregnation of a strawberry with glucose to depress its a_w yields structural changes that differ depending on blanching and incorporation of a co-adjutant to improve texture. The use of calcium lactate as a co-adjutant to improve texture seems to help maintain the cell's original shape and partially avoid the cellular collapse that appears when reducing the fruit's a_w. These microstructural changes correlate well with the texture changes of a strawberry in each of the process stages, as can be seen in Table 10.5.

In general, the developed products have good sensory acceptability, considering their textural changes. The structural changes can be desirable, especially when the developed HMFP will be used as raw material for dehydrated or frozen fruits. Also, it has been demonstrated that blanching and sugar impregnation pre-treatments during HMFP processing favor drying kinetics and the quality of the final product in the case of strawberries (Alvarez et al., 1995), apples (Vergara et al., 1997), and mangoes (Welti-Chanes et al., 1995).

TABLE 10.5. Maximum Extrusion Force for Strawberries with Different Treatments.

Treatment	F (kg)
Fresh	11.5
Steam blanched	9.4
Water blanched	6.2
Glucose impregnated *to* a_w 0.93*	11.4
Glucose impregnated to a_w 0.95*	12.2
Frozen-thawed	5.2
Canned	1.8

* Water activity (a_w) values after attaining equilibrium.

Fruits and vegetables improve the human diet by enriching it nutritionally and organoleptically. Nutritional contributions include considerable amounts of vitamins, minerals, starch, and sugars, which are important sources of dietary and crude fiber. Organoleptically, they are appreciated for their flavor, aroma, texture, and color (Floros, 1993). The nutritional value of fruits has not been of primary concern to producers. However, most research indicates that nutrients are less susceptible to destruction than sensory attributes. Thus, conditions that preserve the physical integrity of fruits and maintain desirable sensory characteristics may also preserve nutrients (Klein, 1987). Average ascorbic acid contents for high moisture papaya and pineapple and their syrups during storage are given in Table 10.6. It can be observed that vitamin loss was not only due to degradation reactions, but also to diffusion in the syrup, and was more significant at 25°C. Alzamora et al. (1989) reported that in pineapple preserved by combined methods after four months of storage at 27°C, only ascorbic acid was present in the syrup. The major ascorbic acid loss in fruit occurs during processing and the first month of storage (Table 10.6). Heng et al. (1990) reported that during the osmotic dehydration of papaya at temperatures between 50°C and 70°C with sucrose syrups (45° to 72° Brix), the ascorbic acid content dropped 40% to 80% during the first 2 hours of treatment. These authors explain the degradation kinetics by the combination of a diffusional phenomenon and chemical degradation. Similar observations were made on the osmotic dehydration of kiwi (Vial et al., 1991) at low temperature where the diffusional phenomenon is the predominant cause of ascorbic acid loss.

The industrial and marketing possibilities of HMFP are linked to the acceptance of the products by the consumer and how nearly the freshness concept has been achieved. Results on the overall acceptability of papaya, peach, pineapple, and mango manufactured as HMFP are presented in Table 10.7 (Alzamora et al., 1993; López-Malo et al., 1994; Corona, 1991). It can be seen that the mean scores indicate good acceptability and that the highest

TABLE 10.6. Ascorbic Acid Content (mg/100 g) in HM Papaya (a_w 0.98, pH 3.5), HM pineapple (a_w 0.97, pH 3.1), and Their Syrups during Storage.

	Papaya			
	5°C		25°C	
Time (Days)	Fruit	Syrup	Fruit	Syrup
Fresh	46.4			
0	23.1	*	23.1	*
15	18.4	13.5	12.0	15.8.
29	9.3	13.0	9.1	12.8
60	8.9	12.2	8.7	12.7
75	7.1	11.9	6.7	9.4
90	6.5	7.7	5.8	6.9

	Pineapple	
	27°C	
Time (Days)	Fruit	Syrup
Fresh	15.35	0
2	6.34	*
13	5.64	4.31
63	4.02	3.20
83	3.38	2.82
100	2.98	2.54
128	0	2.54

* Not determined.

TABLE 10.7. Overall Acceptability of Shelf-Stable High Moisture Papaya, Peach, Pineapple, and Mango

Attribute	Average Score*
Flavor	6.65–7.70
Odor	5.80–6.80
Texture	6.70–8.07
Color	6.46–7.10
Overall impression	6.73–7.63

* Results obtained by untrained panels (30 to 50 members) using a nine-point hedonic scale (9, most acceptable; 1, least acceptable at different storage times (30 to 152 days) and temperatures (25 to 35°C).

values are for texture, so the application of hurdle technology to obtain HMFP has great possibility for succeeding within the MP food market.

The results of the sensory evaluation for some HMFP during storage (López-Malo et al., 1994; Argaiz et al., 1993; Torregiani et al., 1987; Alzamora et al., 1989; Alzamora et al., 1993; Cano et al., 1994; Díaz de Tablante et al., 1993) show that the time needed to reach scores corresponding to well-accepted products varies according to storage conditions and the type of package employed. At 25°C, the shelf life of the products is five to six months (as determined by the color and flavor of the product).

FINAL REMARKS

For a better and more efficient use of hurdle technology in HMFP development, research is needed on the following topics: (1) the mechanism of action for traditional and emerging preservation factors on microorganisms, enzymes, and deteriorative reactions; (2) induced tolerance of pathogenic and spoilage organisms to stress factors; and (3) chemical composition and structural organization of fruit tissues and their changes due to processing factors. Results from these areas will help to identify key factors as to their combined effect on product safety, stability, and quality.

ACKNOWLEDGEMENTS

The authors acknowledge the financial support of the European Project STD-3, the CYTED Project, the Universidad de las Américas-Puebla (Mexico), the Universidad de Buenos Aires (Argentina), and the Universidad Central de Venezuela.

REFERENCES

Ahvenainen, R. 1996. New approaches in improving shelf life of minimally processed fruit and vegetables. *Trends Food Sci. Tech.* 7 (6):179–187.

Alvarez, C. A., Aguerre, R., Gómez, R., Vidales, S., Alzamora, S. M., and Gerschenson, L. N. 1995. Air dehydration of strawberries: Effect of blanching and osmotic pretreatment in the kinetics of moisture transport. *J. Food Eng.* 25 (2):167–178.

Alzamora, S. M., Cerruti, P., Guerrero, S., and López-Malo, A. 1995. Minimally processed foods by combined methods. In *Fundamentals and Applications of Food Preservation by Moisture Control. ISOPOW Practicum II*, ed. G. Barbosa-Cánovas and J. Welti-Chanes, pp. 463–492. Lancaster, PA: Technomic Publishing, Co., Inc.

Alzamora, S. M., Gerschenson, L. N., Cerruti, P., and Rojas, A. M. 1989. Shelf-stable pineapple for long-term non-refrigerated storage. *Lebensm. Wiss. Technol.* 22:233–236.

Alzamora, S. M., Gerschenson, L. N., Vidales, S. L., and Nieto, A. 1997. Structural changes in the minimal processing of fruits: Some effects of blanching and sugar impregnation. In *Food Engineering 2000*, ed. P. Fito, E. Ortega-Rodriguez, and G. V. Barbosa-Cánovas, pp. 117–140. New York: Chapman & Hall.

Alzamora, S. M., Tapia, M. S., Argaiz, A., and Welti, J. 1993. Application of combined methods technology in minimally processed foods. *Food Res. Int.* 26:125–130.

Argaiz, A., López-Malo, A., and Welti, J. 1995. Considerations for the development and the stability of high moisture fruit products during storage. In *Fundamentals and Applications of Food Preservation by Moisture Control. ISOPOW Practicum II*, ed. G. Barbosa-Cánovas and J. Welti-Chanes, pp. 729–760. Lancaster, PA: Technomic Publishing Co., Inc.

Argaiz, A., Vergara, F., Welti, J., and López-Malo, A. 1993. Durazno conservado por factores combinados. In *Boletín Internacional de Divulgación No. 1 del Proyecto CYTED: Preservación de Frutas a Granel por el Método de los Factores Combinados*, pp. 22–30. Universidad de las Américas-Puebla, Mexico.

Cano, M. E., López-Malo, A., and Argaiz, A. 1994. Mango and papaya slices minimally processed by combined methods. In *Proceedings of the Poster Session of ISOPOW Practicum II: Food Preservation by Moisture Control*, ed. A. Argaiz, A. López-Malo, E. Palou, and P. Corte, pp. 42–45. Universidad de las Américas-Puebla, Mexico.

Cerruti, P., and Alzamora, S. M. 1996. Inhibitory effects of vanillin on some food spoilage yeasts in laboratory media and fruit purées. *Int. J. Food Microbiol.* 29:379–386.

Cerruti, P., Alzamora, S. M., and Chirife, J. 1988. Effect of potassium sorbate and sodium bisulfite on thermal inactivation of *Saccharomyces cerevisiae* in media of lowered water activity. *J. Food Sci.* 53:1911–1912.

Cerruti, P., Alzamora, S. M., and Chirife, J. 1990. A multiparameter approach to control the growth of *Saccharomyces cerevisiae* in laboratory media. *J. Food Sci.* 55:837–840.

Cerruti, P., Alzamora, S. M., and Vidales, S. L. 1997. Vanillin as an antimicrobial for producing shelf-stable strawberry purée. *J. Food Sci.* 62:1–3.

Consuegra, R., Tapia de Daza, M. S., López-Malo, A., Corte, P., and Welti-Chanes, J. 1997. Unpublished manuscript, Dept. of Chemical and Food Engineering. Universidad de las Américas-Puebla, Ex-Hacienda Santa Catarina Mártir, Cholula, Puebla, Mexico.

Corona, R. 1991. Evaluación microbiológica del proceso de elaboración y estabilidad de higos (*Ficus carica* L.) conservados por métodos combinados. M.S. thesis, Universidad Central de Venezuela, Caracas, Venezuela.

Corte, P., Díaz, R. V., Tablante, A., Argaiz, A., and López-Malo, A. 1994. Ensayos de reto microbiano con *Zygosaccharomyces rouxii* y *Aspergillus ochraceus* en frutas conservadas mediante la technología de obstáculos. In *Boletín Internacional de Divulgación No. 2 del Proyecto CYTED: Preservación de Frutas a Granel por el Método de los Factores Combinados*, pp. 24–29. Universidad de las Américas-Puebla, Mexico.

Dave, B. P. P., and Gorns, C. G. M. 1993. Modified atmosphere packaging of fresh produce on the West-European market. *ZFL Intç Z. Lebensm.-Tech. Mark. Verpack. Anal.* 44 (1/2):32–37.

Díaz de Tablante, R. V., Tapia de Daza, M. S., Montenegro, G., and González, I. 1993. Desarrollo de productos de mango y papaya de alta humedad estabilizados por métodos combinados. In *Boletín Internacional de Divulgación CYTED No. 1*, pp. 5–21. Universidad de las Américas-Puebla, Mexico.

Dorantes-Alvarez, L. 1997. El oscurecimiento enzimático en el aguacate (*Persea americana*). Ph.D. dissertation. Universidad Politécnica de Valencia, Spain.

Fito, P., and Chiralt, A. 1995. An update on vacuum osmotic dehydration. In *Fundamentals and Applications of Food Preservation by Moisture Control. ISOPOW Practicum II*, ed. G. Barbosa-Cánovas and J. Welti-Chanes, pp. 351–374. Lancaster, PA: Technomic Publishing Co., Inc.

Fito, P., and Chiralt, A. 1997. Osmotic dehydration: An approach to the modeling of solid-liquid food operations. In *Food Engineering 2000*, ed. P. Fito, E. Ortega-Rodríguez, and G. V. Barbosa-Cánovas, pp. 231–252. New York: International Thomson Publishing.

Floros, J. D. 1993. The shelf life of fruits and vegetables. In *Shelf Life Studies of Foods and Beverages. Chemical, Biological, Physical and Nutritional Aspects*, ed. G. Charalambous, pp. 195–216. Amsterdam: Elsevier Science Publishers.

Gould, G. W. 1989. *Mechanisms of Action of Food Preservation Procedures.* Essex: Elsevier Science, Publishers.

Guerrero, S., Alzamora, S. M., and Gerschenson, L. N. 1996. Optimization of a combined factors technology for preserving banana purée to minimize color changes using response surface methodology. *J. Food Eng.* 28:307–322.

Heng, K., Guilbert, S., and Cuq, J. L. 1990. Osmotic dehydration of papaya: Influence of process variables on the product quality. *Sciences des Aliments.* 10:831.

Huxsoll, C. C., and Bollin, H. R. 1989. Processing and distribution alternatives for minimally processed fruits and vegetables. *Food Technol.* 43 (2):132.

Klein, B. P. 1987. Nutritional consequences of minimal processing of fruits and vegetables. *J. Food Quality* 10:179–193.

Leistner, L. 1985. Hurdle technology applied to meat products of the shelf stable and intermediate moisture food types. In *Properties of Water in Foods in Relation to Quality and Stability,* ed. R. Simatos and J. L. Multon, pp. 309–329. Martinus Nijhoff, Dordrecht, Netherlands.

Leistner, L. 1992. Food preservation by combined methods. *Food Res. Int.* 25:51.

López-Malo, A., Alzamora, S. M., and Argaiz, A. 1995. Effect of natural vanillin on germination time and radial growth of moulds in fruit-based agar systems. *Food Microb.* 12:213–219.

López-Malo, A., Alzamora, S. M., and Argaiz, A. 1997. Effect of vanillin concentration, pH and incubation temperature on *Aspergillus flavus, Aspergillus niger, Aspergillus ochraceus,* and *Aspergillus parasiticus* growth. *Food Microbiol.* 14:117–124.

López-Malo, A., Parra, L., and Argaiz, A. 1992. Individual and combined effects of water activity, pH, and potassium sorbate concentration on the growth of three molds. Presented at *IFTEC,* The Hague, Netherlands. Nov. 15–18.

López-Malo, A., Palou, E., Welti, J., Corte, P., and Argaiz, A. 1994. Shelf-stable high moisture papaya minimally processed by combined methods. *Food Res. Intl.* 27:545–553.

Mazzota, A., Ipina, S., Huergo, M., and Alzamora, S. M. 1990. Influencia de la a_w y el pH en el crecimiento y supervivencia de *Bacillus coagulans.* Presented at *XIX Congreso Latinoamericano de Química.* Buenos Aires, Argentina, Nov. 5–9.

Mertens, B., and Knorr, D. 1992. Developments of nonthermal processes for food preservation. *Food Technol.* 46 (5):124–133.

Monsalve-González, A., Barbosa-Cánovas, G., Cavalieri, R. P., McEvily, A., and Iyengar, R. 1993. Control of browning during storage of apple slices preserved by combined methods. *J. Food Sci.* 58:797–800, 826.

Ohlsson, T. 1994. Minimal processing-preservation methods of the future: An overview. *Trends Food Sci. Technol.* 5:341–344.

Palou, E., López-Malo, A., Barbosa-Cánovas, G. V., Welti-Chanes, J., and Swanson, B. G. 1997. High hydrostatic pressure as a hurdle for *Zygosaccharomyces bailii* inactivation. *J. Food Sci.* 62 (4):855–857.

Resurrección, A. V. A., and Prussia, S. E. 1986. Food related attitudes: Differences between employed and unemployed women. In *Proceedings of the 32nd Annual Conference of the American Council on Consumer Interests,* American Council on Consumer Interests p. 156. Columbia, MO.

Rolle, R. S., and Chism, G. W. 1987. Physiological consequences of minimally processed fruits and vegetables. *J. Food Qual.* 10:187.

Sánchez-Pardo, M. E., Ortíz-Moreno, A., and Dorantes-Alvarez, L. 1991. The effect of ethylene diamine tetracetic acid on preserving the color of an avocado purée. *J. Food Process. Preserv.* 15:261–271.

Shewfelt, R. 1987. Quality of minimally processed fruits and vegetables. *J. Food Qual.* 10:143–156.

Tapia de Daza, M. S., Alzamora, S. M., and Welti-Chanes, J. 1996. Combination of preservation factors applied to minimal processing of foods. *Critical Rev. Food Sci. Nutrition.* 36 (6):629–659.

Tapia de Daza, M. S., Argaiz, A., López-Malo, A., and Díaz, R. V. 1995. Microbial stability assessment in high and intermediate moisture foods. Special emphasis on fruit products. In *Fundamentals and Applications of Food Preservation by Moisture Control. ISOPOW Practicum II,* ed. G. Barbosa-Cánovas and J. Welti-Chanes, pp. 575–602. Lancaster, PA: Technomic Publishing Co., Inc.

Torregiani, D., Forni, E., and Rizzolo, A. 1987. Osmotic dehydration of fruit. Part 2: Influence of the osmosis time on the stability of processed cherries. *J. Food Proc. Preserv.* 12:27–44.

Vergara, F., Amézaga, E., Bárcenas, M. E., and Welti-Chanes, J. 1997. Analysis of the drying processes of osmotically dehydrated apple using the characteristic curve model. *Drying Tech.* 15 (3&4):949–963.

Vial, C., Guilbert, S., and Cuq, J. L. 1991. Osmotic dehydration of kiwi fruits: Influence of process variables on the quality. *Sciences des Alimentes.* 11:63.

Vidales, S. L., Castro, M. A., and Alzamora, S. M. 1994. Conservación de frutilla: Influencia del proceso sobre la textura y la pared celular. Presented at *VI Congreso Argentino de Ciencia y Tecnología de Alimentos y 1er Encuentro de los Técnicos de Alimentos del Cono Sur,* Buenos Aires, Argentina. April 6–9.

Welti-Chanes, J., and Vergara, F. 1995. Fruit preservation by combined methods: An Ibero-American research project. In *Fundamentals and Applications of Food Preservation by Moisture Control. ISOPOW Practicum II,* ed. G. Barbosa-Cánovas and J. Welti-Chanes, pp. 444–462. Lancaster, PA: Technomic Publishing Co., Inc.

Welti-Chanes, J., Palou, E., López-Malo, A., and Balseira, A. 1995. Osmotic concentration-drying of mango slices. *Drying Tech.* 13 (1&2):405–416.

Wiley, R. C. 1994. Introduction to minimally processed refrigerated fruits and vegetables. In *Minimally Processed Refrigerated Fruits and Vegetables,* ed. R. C. Wiley. New York: Chapman & Hall.

CHAPTER 11

Minimal Processing of Foods with Thermal Methods

THOMAS OHLSSON

THE THERMAL HEATING APPROACH TO MINIMAL PROCESSING

T HERMAL methods are extensively used for the preservation of foods. Thermal treatments lead to desirable changes, such as protein coagulation, starch swelling, textural softening, and formation of aroma components. However, undesirable changes also occur, such as losses of vitamins and minerals, formation of thermal reaction components of biopolymers, and in minimal processing terms, losses of fresh appearance, flavor, and texture.

The classical approach to overcome or at least minimize these undesirable quality changes in thermal processing is the HTST (high temperature short time) concept. HTST is based on the fact that the inactivation of microorganisms is primarily dependent on the time of the heat treatment, whereas quality changes are primarily dependent on the time duration of the heat treatment (Ohlsson, 1980). High temperatures will give rapid inactivation of microorganisms and enzymes, which are goals of pasteurization or sterilization, and short times will give fewer undesired quality changes. The problem in applying this principle to solid and high-viscosity foods is that the parts of the food in contact with the hot surfaces will be overheated during the time needed for the heat to transfer to the interior or coldest part of the food. The surface overheating will give quality losses in severe cases due to the low heat diffusivity of foods. Thus, direct volume heating methods are seen as minimal processing methods, where the thermal processing is applied to minimize the quality changes of the process (Ohlsson, 1994).

141

Another quality aspect of thermal processes is that juice losses in meat and fish are strongly dependent on the temperatures reached. Only a few degrees will give large differences in juice losses, which is important both to the economic yield of the process and to the consumer-perceived "juiciness" of the product. Avoiding temperatures exceeding the desired culinary and bacteriologically determined final core temperature is therefore important. Consequently, in these cases, thermal processing is best achieved with the LTLT concept, i.e., using low temperatures over long times.

ELECTRIC VOLUME HEATING METHODS
FOR FOODS

In the food industry, thermal processing using electric heating is done by applying electromagnetic energy for producing temperature increases in foods. These temperature increases in turn cause desired changes in the food (e.g., inactivation of microorganisms and enzymes or production of the desired flavor and texture associated with ready-made foods). Many of the electric heating methods are not in themselves new, but the knowledge about them, as well as their application in the food industry, is limited at present (UIE, 1996).

Electric heating methods directly transfer energy from an electromagnetic source to the food without heating transfer surfaces in the heat processing equipment. This direct energy transfer is of major advantage, as it gives excellent opportunities for high energy utilization. Looking through the electromagnetic spectrum, we can identify three frequency areas that are employed today in the industry for direct heating of food. The area of 50 to 60 Hz (equivalent to the electric power in a household) is used for electric resistant heating, sometimes called ohmic heating. In this application, the food itself acts as a conductor between a ground and the charged electrode, normally at 220 or 380 volts. In the high frequency area of 10 to 60 MHz, foods are placed between electrodes (one of them again being grounded), and energy is transferred to the food placed between the electrodes. In the microwave region of 1 to 3 GHz, energy is transferred to the food through the air by guided waves controlled by electromagnetic devices called applicators. In all of these electric heating methods, it is important to have an understanding of the interaction between the electromagnetic field in the frequency in question and of the material being subjected to the energy. Electric and dielectric properties of foods and other materials in construction equipment are important to know in order to better understand and control the application of electric energy for the heating of foods (Ohlsson, 1987).

ELECTRIC RESISTANCE/OHMIC HEATING

FUNDAMENTALS

In electric resistance heating, the food itself acts as a conductor of electricity, taken from the mains that are 50 Hz in Europe and 60 Hz in the United States. The food may also be immersed in a conducting liquid, normally a weak salt solution of similar conductivity to the food. Heating is accomplished according to Ohm's law, where the conductivity (i.e., the inverse and/or resistivity) of the food will determine the current that will pass between the ground and electrode. Normally, voltages up to 5,000 V are applied. The conductivity of foods increases considerably with increasing temperature; to reach high temperatures, it is therefore necessary to increase the current of the voltage or to use longer distances between the electrodes and the ground.

EQUIPMENT

The best-known electric resistance heating system is the APV ohmic heating column, where electrodes are immersed into a food that is transported in a vertical concentric tube. Electrodes (usually four) are connected to the earth and to a line voltage. The inside of the tube and electrodes are lined with high temperature, electrically inert plastic material. As the electrodes are often highly conductive metals with undesirable corresponding ions, the isolation of electrodes against the food components is of major importance.

APPLICATIONS

The ohmic system of APV has been installed for pasteurization and sterilization of a number of food products with resulting excellent quality. The majority of these installations are found in Japan for the production of fruit products (Tempest, 1996). Other installations include prepared food in the UK. The ohmic heating system shows excellent retention of particle integrity due to the absence of mechanical agitation, as is typical for traditional heat exchanger-based heating systems. A special reciprocal piston pump is used to accomplish the high particle integrity with a long traditional tubular heat exchanger for the cooling.

Extensive microbiological evaluation of the system has shown that the method can produce sterile products reliably.

Other industrial cooking operations for electric resistance heating involve rapid cooking of potatoes and vegetables for industrial blanching and preparation of foods in institutional kitchens. One of the major problems with these applications is ensuring that the electrode materials are inert and do not release metal ions into the conducting solutions and eventually into the foods. Another problem is the need to properly control the electric conductivity of all the constituents of the food product, as this determines the rate of heating for the different constituents. This often requires well-controlled pre-treatments to eliminate air in food and to control salt levels in foods and sauces (Zoltai and Swearingen, 1996).

HIGH-FREQUENCY HEATING

FUNDAMENTALS

High-frequency heating is conducted in the MHz part of the electromagnetic spectrum. The frequencies of 13.56 and 27.12 MHz designate for industrial heating applications. Foods are heated by transmitting electromagnetic energy through the food placed between an electrode and the ground. The high-frequency energy used will allow for the transfer of energy over air gaps and through non-conducting packaging materials. To achieve sufficient rapid heating in foods, high electric field strengths are needed.

High-frequency heating is accomplished by a combination of dipole heating, when the water dipole tries to align itself with the alternating electric field, and electric resistance heating from the movement of the dissolved ions of the foods. In the lower temperature range (including temperatures below the freezing point of food), dielectric heating is important, whereas, for elevated temperatures, electric conductivity heating dominates. The conductivity loss or dielectric loss factor increases with increasing temperature, which may lead to problems of runaway heating when already hot parts of the food absorb a majority of the supplied energy. Dielectric properties of foods are reasonably abundant in the low temperatures above normal room temperature.

EQUIPMENT

An important part of high-frequency heating equipment is the design of the electrodes. A number of different configurations depend on the field strength needed and/or the configuration of the sample. For high moisture applications, the traditional electrode configuration is most often used. For low moisture applications, such as dried foods and biscuits, electrodes in the form of rods that give stray fields for food placed on a conveyor belt are often used. Because electrodes can be designed to create uniform electric field patterns and heating patterns for different types of food geometries, they can now be supported by computer simulation techniques, such as FEM software packages (Metaxas, 1996).

High-frequency power is generated in a circuit. The circuit consists of a coil, condensator plates with the food in between, an amplifier in the form of a triode, and an energy source. Modern electronic control devices are also employed to maintain a given frequency, as this may vary as the food is heated (bearing in mind that the food itself is part of the oscillating circuit). The control function has recently been improved by the introduction of so-called 50 ohms technology, which allows a separate control for tuning the load circuit.

APPLICATIONS

The largest application in the food industry for high-frequency heating is in the finish drying or postbaking of biscuits and other cereal products. Another application is the drying of products, such as expanded cereals and potato strips. Previously, defrosting frozen foods using high frequencies was a major application, but problems of uniformity with foods of mixed composition have limited actual use. Nevertheless, interest in high-frequency defrosting has increased again in the past few years.

Recently, high-frequency cooking equipment for pumpable foods has been developed. These devices involve pumping a food through a plastic tube shaped to give uniform heating when placed between two electrodes. Excellent temperature uniformity has been demonstrated in these applications, especially for the continuous cooking of ham and sausage emulsions (Tempest, 1996).

MICROWAVE HEATING

FUNDAMENTALS

Microwaves used in the food industry for heating utilize ISM (Industrial, Scientific and Medical) frequencies of 2,450 MHz or 915 MHz, which correspond to 12 or 34 cm in wavelength. In this frequency range, the dielectric heating mechanism dominates up to moderate temperatures. Polar molecules try to align themselves to the rapidly changing direction of the electric field. This alignment requires energy that is taken from the electric field. When the field changes direction, the molecule relaxes, and the energy previously absorbed is dissipated to the surroundings directly inside the food. The penetration depth from one side is approximately 1 to 2 cm at 2,450 MHz. At higher temperatures, electric resistance heating from dissolved ions also plays a role in the heating mechanisms, further reducing the penetration depth of the microwave energy. The limited penetration depth of microwaves implies that the distribution of energy within the food can vary. The control of microwave heating uniformity is difficult, as the objects to be heated are of the same size as the wavelength in the material. The difficulties in controlling heating uniformity must be seen as the major limitation of the industrial application

TABLE 11.1. **Factors Influencing Microwave Heating Uniformity.**

- Food composition and geometry
- Packaging geometry and composition
- Microwave energy feed system

of microwave heating. Thus, an important requirement for microwave equipment and microwave energy application in the food industry is the ability to properly control heating uniformity (Ohlsson, 1983).

EQUIPMENT

The transfer of microwave energy to foods occurs through contactless wave transmission. The microwave energy feed system is designed to control uniformity during the heating operation. Many different designs are used in industrial applications, from the traditional multimode cavity oven via direct radiation waveguide applicators, through sophisticated periodic structures (Metaxas, 1996). The interaction of parameters important to heating uniformity must be taken into account when designing applicators (Table 11.1).

As pointed out by Ryynänen and Olhsson (1996), the importance of food geometry and the actual layout of the components on a plate in order to reach good heating uniformity is often poorly understood.

The microwave energy feed system controls electric field polarization, which in turn is responsible for overheating of food edges. This is one of the most severe causes of uneven microwave heating in foods (Sundberg et al., 1996).

The very high frequencies used in microwave heating allow for rapid energy transfers and, thus, high rates of heating. This is a major advantage, but can also lead to problems of non-uniform heating when excessively high energy transfer rates are used.

APPLICATIONS

Industrial applications of microwave heating are found for most of the heat treatment operations in the food processing industries. For many years, the largest application has been defrosting or thawing frozen foods, such as blocks of meat prior to further processing. Often, meat is only partially defrosted (tempered) before it can be further processed. Another large application area is pasteurization, and now sterilization, of packaged foods. Primarily, ready-made foods are those that are processed. The objective of these operations is to pasteurize foods to temperatures in the range of 75°C to 80°C in order to prolong their shelf life to approximately three to four weeks. Sterilization using microwaves has been investigated for many years, but commercial introduction has only come in the past few years in Europe and Japan. Microwave pasteurization and sterilization promise to give very quick heat processing, which should lead to small quality changes due to thermal treatments according to the HTST principle. However, it has turned out that very high requirements for heating uniformity must be met in order to fulfill these quality advantages (Ohlsson, 1991).

Pasteurization with microwave heating can also be done with pumpable foods, as pointed out by Püschner (1964). Microwaves are directed to the tube where the food is transported, and heating is accomplished directly across the tube cross section. Again, uniformity of heating must be ensured, which requires selection of the correct dimensions of the tube diameter and proper design of the applicators (Ohlsson, 1993). Systems where food is transported through the heating zone by a screw are also available (Berteaud, 1995).

Another application of microwave heating is drying in combination with conventional hot air drying. Microwaves are primarily used for moving water from the wet interiors of solid food items to the surface, relying on the preferential heating of water by microwaves. Applications can be found for pasta, vegetables, and various cereal products where puffing by rapid expansion of the interior of the food matrix can be accomplished using microwave energy (Tempest, 1996).

Microwave energy is also used for various cooking and coagulation processes for meat products, chicken, and fish, often in combination with other conventional cooking operations. A number of new applications in the microwave heating area have been reported recently, often involving the utilization of unique microwave properties of higher energy fluxes and direct interior heating.

COMPARING THE FREQUENCIES

The advantages and limitations for each of the various frequencies are listed in Table 11.2.

TABLE 11.2. Comparisons between the Frequencies of Electric Heating.

Ohmic and High Frequency	Microwave
+ Better for large, thick foods + Lower investment costs + Easier to understand and control	+ Higher heating rate + Design freedom + Less sensitive to food inhomogenity + Much R&D available
− Risk of arching in HF − Larger floor space − Narrow frequency bands − Limited R&D support	− Limited penetration − Higher investment costs − More engineering needed

SAFETY ASPECTS

In the application of electromagnetic energy for the heating of foods, questions about safety in terms of nutritional value and the existence of nonthermal heating effects have often been raised. From extensive studies regarding the changes of chemical constituents in foods as a result of electric heating, it has been demonstrated that the effect of electromagnetic heating in all practical aspects is the same as for conventional heating to the same temperature.

Electric heating equipment for the food industry must be designed and operated according to international and national safety standards. The levels of allowable leakage vary over the frequency range according to these standards. Equipment for measuring and monitoring electromagnetic energy leakage from electric equipment is readily available (IEC, 1982).

REFERENCES

Berteaud, A.-J. 1995. *Thermo-Star.* Bulletin from MES, 15, Rue des Solets, RUNGIS, France.

IEC, 1982. *Safety in Electroheat Installations.* Publication 519. IEC, Geneva, Switzerland.

Metaxas, A. C. 1996. *Foundation of Electroheat: A Unified Approach.* Chichester, UK: J. Wiley and Sons Ltd.

Ohlsson, T. 1980. Temperature dependence of sensory quality changes during thermal processing. *J. Food Sci.* 45 (4):836–839.

Ohlsson, T. 1983. Fundamentals of microwave cooking. *Microwave World* 4 (2):4–9.

Ohlsson, T. 1987. Dielectric properties: Industrial use. In *Physical Properties of Foods—2.* London: Elsevier Applied Science Publ. pp. 199–212.

Ohlsson, T. 1991. Microwave processing in the food industry. *Euro Food Drink Rev.* 3–6.

Ohlsson, T. 1993. In-flow microwave heating of pumpable foods. Presented at the *International Congress on Food and Engineering,* Chiba, Japan, May 23–27, p. 7.

Ohlsson, T. 1994. Minimal processing: Preservation methods of the future: an overview. *Trends in Food Sci. Technol.* 5 (11):341–344.

Püschner, H. 1964. *Wärme durch Mikrowellen.* Endhoren, Holland: Philips Techn. Bibliotek.

Ryynänen, S., and Ohlsson, T. 1996. Microwave heating uniformity of ready meals as affected by placement, composition, and geometry. *J. Food Sci.* 61 (3):620–624. SIK-Publication No. 735.

Sundberg, M., Risman, P. O., Kildal, P.-S., and Ohlsson, T. 1996. Analysis and design of industrial microwave ovens using the finite difference time domain method. *J. Microwave Power Electromag. Energy.* 31 (3):142–157.

Tempest, P. 1996. *Electroheat Technologies for Food Processing.* Bulletin of APV Processed Food Sector, Crawley, W. Sussex, England.

UIE. 1996. *Electricity in the Food and Drinks Industry.* UIE, B.P. 10. Place de la Defense, Paris, France.

Zoltai, P., and Swearingen, P. 1996. Product development considerations for ohmic heating. *Food Technol.* 50 (5):263.

CHAPTER 12

Blanching of Fruits and Vegetables Using Microwaves

LIDIA DORANTES-ALVAREZ
GUSTAVO V. BARBOSA-CÁNOVAS
GUSTAVO GUTIÉRREZ-LÓPEZ

INTRODUCTION

U SE of microwaves to thermally treated foodstuffs began in 1949 when Percy Spencer realized that radar waves could warm food products. He patented his idea in 1952. However, it wasn't until the development of the microwave oven that this technique proved to be a success on a major scale.

Microwaves consist of electromagnetic energy, with wavelengths between 2.5 and 75 cm in the electromagnetic spectrum, which is located between radio waves and infrared radiation. Under vacuum, microwaves propagate at the velocity of light; through other media, their speed decreases according to the change of the refractive index. Similar to all electromagnetic waves, microwaves present characteristics of reflection, absorption, and transmission when interacting with different media.

Factors that influence the heating velocity in microwave ovens are the type of magnetron, power supply, load to the oven, geometry of the product, thermal and dielectric characteristics of the food, initial temperature of the sample, specific heat of the containing vessel, and position of the load in the chamber (Buffler, 1993). Rosen (1972) described the electromagnetic spectra according to the frequencies associated with each type of radiation of specific wavelength, as well as the different types of chemical bonds that specific radiation may break. According to these data, microwaves themselves are not able to break covalent chemical links. However, due to the oscillatory kinetic behavior of molecules when subjected to microwaving, it is reasonable to think that they may induce changes that are hard to evaluate, due to a complex combined

149

thermal and mechanical effect, making the isolated effect of microwaves on the foodstuff difficult to assess.

The rate of heating when using microwaves is the main advantage of this technique. Also, because microwaves penetrate the sample, heating is accomplished in the interior of the food. When heated rapidly, the quality of fruits and vegetables such as flavor, texture, color, and vitamin content is better kept. In Table 12.1, selected works on blanching by microwaving foodstuffs are presented. The selection does not pretend to be comprehensive, but rather illustrative of relevant research in this area. Comments on blanching conditions, as well as remarks on the effects on quality are also included. The authors underline the advantages of microwaving over immersion in boiling water or steaming. For example, blanching spinach with microwaves is cited as a better choice than traditional methods for retention of vitamin C. This may be due to the faster processing when using microwaves (Quenzer and Burns 1981). Esaka et al. (1987) reported that during the blanching of beans, trypsin inhibitors as well as lipoxygenase were more rapidly inactivated when using microwaves. Torringa et al. (1992) subjected mushrooms to blanching by microwaves and obtained a product with better aroma and texture compared to traditionally treated samples.

In this chapter, original results obtained by the authors are also included about the microwave blanching of avocado paste in comparison to traditional methods. In the case of avocado, development of off-flavors during conventional thermal processing represents serious limitations for marketing, as first reported by Bates (1970). However, because enzymes such as polyphenoloxidase (PPO EC: 1.10.3.1) remain active in nonthermally processed products and cause browning, their inactivation must be accomplished under an alternative process. It is necessary to investigate processing conditions for the thermal inactivation of PPO that produce a minimum flavor deterioration. Muftugil (1986) compared four methods for blanching green beans and found that the time for peroxidase inactivation was lower using microwaves compared to conventional methods. Some authors (Khalil and Villota, 1988; Tajchakavit and Ramaswamy, 1996) have made attempts to show the athermal effects associated with microwave heating, but Buffler (1993) has been cautious about the presence of nonthermal effects, stating that, by microwave heating, only an enhanced energy transmission is achieved. It is believed that the measurement of food temperature in a microwave oven can give more information about this controversial hypothesis. Given this situation, the purpose of this study was to find and evaluate any differences in the inactivation characteristics of avocado PPO by microwave and conventional heating processes.

MATERIALS AND METHODS

Avocado fruits (*Persea americana,* Mill.) of the Haas variety were obtained

TABLE 12.1. Some Experimental Reports on Effects of Microwave Blanching on Fruit and Vegetable Products.

Commodity	Conditions of Microwave Blanching	Effect on Quality	Comments	Reference
Strawberry (concentrate)	4 kg of fruits 70/50 microwave oven 3–4 min	Blanched strawberry browns slower than the control	Blanching has protective effects on pigments, ascorbic acid, and color stability	Wrolstad et al. (1980)
Spinach	100 g sample 0.650 KW oven 1 1/2 min	Texture firm in microwave blanched spinach and a better retention of vitamin C	Microwave blanching was faster than the conventional method	Quenzer and Burns (1981)
Mushrooms (canned)	600 g sample 0.65 RW 4 min 40 sec oven	Color and sensory analysis showed no differences between the two blanching methods	Microwave blanching increased yield compared to traditional blanching	Baldwin et al., (1986)
Winged bean seeds	20 g samples 0.5 RW power oven 2.450 MHz 3.5 min	Microwave blanching increased the rate of water adsorption	Blanching inactivated trypsin inhibitors faster than the traditional method	Esaka et al. (1987)
Potatoes (whole)	600 kg/h 3 KW in a 2,450 MHz tunnel 6 min		Continuous process	IFT Expert panel (1989)
Canola seeds	500 g samples 2450 MHz 0.31–1.5 RW 30–630 sec	Microwave treatment increased the yellow color of the seed oils	Microwaves were an alternative to myrosinase inactivation	Owusu-Ansah and Marianchak (1991)
Mushrooms		Microwave-blanced mushrooms had better aroma and texture	By application of microwave energy, an improvement of net yield was achieved	Torringa et al. (1992)

from a local market in Mexico City. Their lipid content was $17 \pm 0.5\%$, and they showed an even black peel color.

PREPARATION OF THE ENZYMATIC EXTRACT

Acetone powder was prepared from avocados by the method described by Dizik and Knapp (1970). Three extractions were performed with acetone. PPO was extracted from the powder by suspending it in distilled water (1:100 w/w) and stirring with a Braun domestic plastic mixer model 4-172 for 1 min; the mixture was centrifuged in a Beckman Ultracentrifuge at 11,000 rpm for 30 min at 10°C. The supernatant used to perform the work contained 0.55 mg protein/ml and was evaluated by the method of Lowry et al. (1959).

MICROWAVE TREATMENT OF THE POLYPHENOLOXIDASE EXTRACT

Samples (5 ml) of the PPO extract placed in 50-ml beakers were heated in a 700-watt Panasonic microwave oven at medium/low power. Processing times in the oven were 10, 20, 30, 35, and 40 seconds. Temperatures were recorded by a Photonetics multisensor (Metricor 1450, Woodensville, MA), including fiberoptic probes that were inserted in the samples. After heating, the samples were cooled in an ice-water bath to room temperature (20 ± 1°C).

CONVENTIONAL HEATING

Five milliliters of the PPO extract were placed in test tubes equipped with a thermocouple and heated by rotation with a gas burner to reach the same temperatures achieved in microwave processing. These samples were then cooled in an ice-water bath to room temperature.

ADDITION OF COPPER IONS

This experiment was carried out to investigate any possible effect due to the addition of copper ions to the enzymatic extract before treatments. Solid cuprous chloride, cupric chloride, cupric carbonate, or cupric acetate was added to 100 ml of PPO preparations to give a final concentration of 10^{-4} Molar. The prepared samples were processed by both microwave and conventional heating as described above.

POLYPHENOLOXIDASE ACTIVITY

The rate of browning caused by the oxidation of catechol catalyzed by polyphenoloxidase was determined by the increased absorbancy of the samples at 420 nm, as described by Halpin and Lee (1987). Enzyme extract (0.5 ml) was then mixed with 3 ml of catechol solution to reach a final concentration of 1% catechol in the sample cuvette. The rate of PPO-catalyzed browning in the catechol solution was monitored at 10- to 180-second intervals in a Shimadzu UV 210 A spectrophotometer. One unit of enzyme activity was defined as a 0.001 change of absorbance $second^{-1}$ (ml of extract)$^{-1}$.

ELECTROPHORETIC PATTERN

Samples of PPO extract were heated as described above and then were analyzed by polyacrylamide gel electrophoresis with 15 μl of each sample containing 7.5 μg of protein applied to the gel. A mini vertical electrophoresis system (GIBCO-BRL, from Gaithersburg, MD) was used, leaving the process to run for 3 hours on six proteins of different molecular weight. They were then applied to another gel to obtain a regression curve of molecular weight versus R_f values. The proteins were developed with Coomasie blue R-250 and PPO activity of samples detected by incubating 30 minutes with 1% catechol.

INACTIVATION OF PPO IN AVOCADO PURÉE

The avocado purée was prepared and then blanched by microwave or conventional heating. The size of the samples was 20 g, which was determined during preliminary experiments as the most convenient to achieve an even temperature distribution. Blanching or PPO inactivation was considered when 1 g of purée on a filter paper impregnated with 0.2 ml of a 2% catechol solution for 20 minutes did not show any developments of chromophore and the percentage of treated avocado puree reflectance as evaluated in an Agtron 400A spectrophotometer was constant for 60 minutes. All the above described experiments were repeated five times.

SENSORY EVALUATION

A test for flavor was carried out with 10 semi-trained panelists selected from 40 people subjected to a sensory-scale test as described by Pedrero and Pangborn (1989). A hedonic test was performed using a 0 to 10 semi-structured scale with the following definitions: 0 to 2.5 dislike, 2.6 to 5.0 like slightly, 5.1 to 7.5 like, 7.6 to 10 like very much.

RESULTS AND DISCUSSION

TEMPERATURE GRADIENT DURING HEATING OF PPO EXTRACTS

Figure 12.1 shows temperature values versus time during the microwave

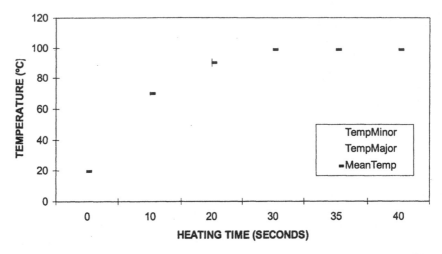

Figure 12.1 Temperatures of polyphenoloxidase extract samples heated conventionally and by microwave.

heating of the PPO extracts. The mean and standard deviation of the temperature values in each case were taken into account to construct the graphs. The temperature increased with the time of heating until it reached 99°C. This heating pattern was reproduced in conventional heating experiments as described previously.

INACTIVATION OF POLYPHENOLOXIDASE IN CRUDE EXTRACT

The residual activity of the conventionally heated and microwave-heated PPO extract is shown in Table 12.2. Constant rates were calculated as the slopes of the initial part of the curves of absorbance versus time and are given in activity units. PPO extract temperatures can also be observed, revealing that there was no significant difference ($P < 0.05$) between the residual activities obtained by both methods of heating at the same temperature. A complete inactivation of PPO crude extract was achieved when temperatures were maintained at 99°C for 10 seconds for either method of heating. These results are similar to those found by Montgomery and Petropakis (1980), who reported that in order to inactivate Bartlett pear polyphenoloxidase with heat in the presence of ascorbic acid, temperatures of up to 90°C must be applied.

ELECTROPHORETIC PATTERN OF HEATED SAMPLES

Figure 12.2 shows the electrophoretic pattern of crude extract avocado PPO samples heated conventionally and with microwaves. Sample 0 corresponds to the blank (no heated extract) and shows one fraction of PPO with a molecular

TABLE 12.2. Residual Activity of Avocado PPO after Heating.

Temperature Reached in the Samples	Microwave Treatment Samples	k Microwave (A.U.)*	Conventional Treatment Samples	k' Conventional (A.U.)*
Not treated	0	45.4 ± 1.4	0	45.4 ± 1.4
69.9 ± 0.9°C	1	43.4 ± 1.8	1'	43.8 ± 1.3
90 ± 2.8°C	2	41.8 ± 2.0	2'	42.2 ± 1.7
99°	3	19.4 ± 3.0	3'	23.4 ± 4.1
99°C (5 sec)		0.72 ± 0.15		0.88 ± 0.16
99°C (10 sec)	4	0	4'	0

*Activity units 1 A.U. = 0.001 ΔA_{420} s^{-1} (ml of extract)$^{-1}$. k Microwave: rate constant obtained by microwave heating; k' conventional: rate constants obtained by conventional heating.

155

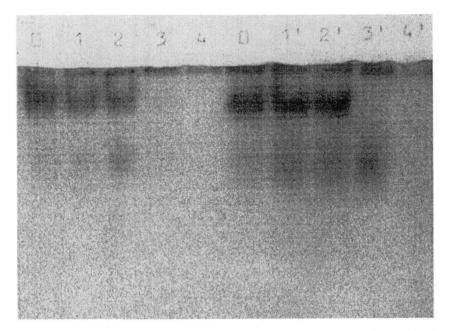

Figure 12.2 Electrophorectic patterns of PPO extracts heated by microwave and conventional methods. Sample 0 was not heated; numbers 1, 2, 3 and 4 correspond to samples heated 10, 20, 30, and 40 seconds in microwave oven. Samples labeled 1′, 2′, 3′ and 4′ were heated by conventional method.

weight of 119,000 and another with that of 59,000. These data are similar to reports in the literature by Dizik and Knapp (1970), who obtained five fractions of crude avocado PPO with estimated molecular weights of 14, 28, 56, 112, and more than 400,000 daltons, and Martinez and Whitaker (1995), who listed the molecular weight of mushrooms polyphenoloxidases at 128,000 with four subunits and that for humans and rats at 62,600 and 58,000, respectively. In the gel, numbers 1, 2, 3, and 4 correspond to the PPO extract samples heated for 10, 20, 30, and 40 seconds in a microwave oven as described in the Methods section (for residual activities see Table 12.2) with those samples labeled 1′, 2′, 3′, and 4′ heated by the conventional method. In both groups, the increasing numbers correspond to an enhanced treatment temperature. As can be seen in Figure 12.2, the fraction of PPO (the first from the top) with the higher molecular weight shows activity in samples 0, 1, 2, 1′, and 2′, which means that when the heating treatment reached temperatures higher than 90°C, this fraction was inactivated (samples 3, 4, 3′, and 4′). On the other hand, the fraction with the lower molecular weight (second from the top) was found to be more heat resistant because it wasn't inactivated until the temperature in the extract prevailed for 10 seconds at 99°C (samples 4 and 4'). Also, the chromophore, which is the product of the reaction of the lower fraction of PPO and catechol, was at a higher concentration in those samples heated at 90°C (2 and 2') than in the unheated sample (0). This suggests that heating helps to unfold the PPO of higher molecular weight, producing lower molecular weight units.

EFFECT OF COPPER IONS

Results on the effect of copper ions (Table 12.3) show that cuprous chloride helps to maintain 1.3% of PPO activity when PPO is denatured by conventional heating and microwaves. Addition of cupric acetate or cupric carbonate is also shown to be effective in maintaining certain PPO activity on denaturation (2.9 and 2.5% of the original activity), but only when the enzyme is treated with microwaves. Cupric chloride was not effective, probably due to the inhibitory effect of chloride ions. The proposed mechanism for the catalysis of PPO described by Whitaker (1994), in which cupric rather than cuprous ions in PPO molecules appear to have a coordination link with water, was considered to explain why cupric ions can maintain certain PPO activity along denaturation by microwave heating. Given the fact that this molecule of water is a dipole, it moved rapidly upon application of microwaves, de-stabilizing the link to the cupric ion and thus helping to release it.

TEMPERATURE GRADIENT DURING HEATING OF AVOCADO PURÉE

Figure 12.3 presents the mean of temperature vs. time during microwave and conventional heating, indicating that the puree samples subjected to conventional heating took longer to reach the temperatures obtained by microwave heating.

INACTIVATION OF PPO IN AVOCADO PURÉE

To inactivate PPO in samples of avocado purée (pH 6.6), treatment at 103°C for 5 seconds was necessary. This was achieved by heating 20 g of purée for 23 seconds at medium/low power in the microwave oven or on the hot plate for 45 seconds (see Figure 12.2). Keeping in mind that the optimum pH for avocado PPO is in the range of 6 to 7, when the pH was lowered to 4.3 using citric acid, the treatment time was lowered to 2 seconds at 103°C and diminished to 1 second at 103°C when the pH was 3.9. These data point out the importance of pH values in the inactivation time of avocado PPO. Aguilera et al. (1987) reported that, in order to obtain 88% inactivation of PPO in Sultana grapes, they should be heated alternatively for 98°C/120 s, 93°C/150 s, or 88°/180 s. The differences in temperature/time of treatments

TABLE 12.3. Effect of Copper Ions on Denatured, Heated PPO Extract Activity.

Microwave Heated PPO Extract	k (A.U.)*	Conventional Heated PPO Extract	k' (A.U.)*
+ CuCl 10^{-4} M	0.64 ± 0.10	+ CuCl 10^{-4} M	0.56 ± 0.10
+ CuCl$_2$ 10^{-4} M	0	+ CuCl$_2$ 10^{-4} M	0
+ Cu (CH$_3$COO)$_2$ 10^{-4} M	1.32 ± 0.24	+ Cu (CH$_3$COO)$_2$ 10^{-4} M	0
+ CuCO$_3$ 10^{-4} M	1.12 ± 0.12	+ CuCH3CO3 10^{-4} M	0
Non-heated PPO extract	45.4 ± 1.4	Non-heated PPO extract	45.4 ± 1.4

*Activity units: 1 A.U. $= 0.001 \Delta A_{420}$ s^{-1} (ml of extract)$^{-1}$. k Microwave: rate constants obtained by microwave heating; k' conventional: rate constants obtained by conventional heating.

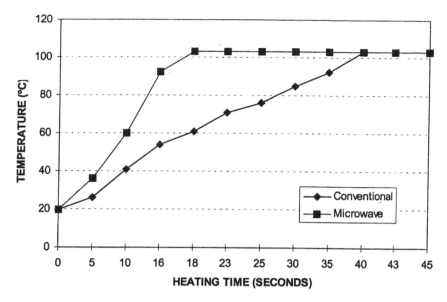

Figure 12.3 Avocado puree temperature changes during microwave or conventional heating.

in that study and the present one can be explained depending upon the characteristics of the fruits and conditions of heating. However, there is a coincidence in the range of temperatures used to obtain significant inactivation of the enzyme in both fruits.

SENSORY EVALUATION OF BLANCHED SAMPLES OF AVOCADO PURÉE

Results on the sensory evaluation of the unheated purée and blanched samples by both methods revealed a significant difference ($P < 0.05$) between the three samples from a two-way variance analysis. No significant difference among panelists was found. A "like very much" qualification was given to the unheated purée, and a "like" mark was given to the heated samples. A significant difference ($P < 0.05$) between the heated samples showed preference toward the microwave blanched purées; this was probably due to a shorter treatment time and, therefore, better retention of flavor. Considering that a rapid increase in temperature can be achieved in a microwave oven that not only inhibits enzymatic browning, but that also preserves flavor, microwave blanching offers a good alternative for PPO inactivation in avocados.

ACKNOWLEDGEMENTS

The authors want to acknowledge the support given by the Instituto

Politécnico Nacional, Dra. Rosalba Mora, and the CYTED Project XI.3 "Desarrollo de tecnologías para la conservación de alimentos por tratamientos mínimos."

REFERENCES

Aguilera, J. M., Oppermann, K., and Sánchez, F. 1987. Kinetics of browning of Sultana grapes. *J. Food Sci.* 52:990–994.

Baldwin, D. R., Anantheswanan, R. C., Sastry, S. K., and Beelman, R. B. 1986. Effect of microwave blanching on the yield and quality of canned mushrooms. *J. Food Sci.* 51:965–966.

Bates, R. P. 1970. Heat induced off flavor in avocado flesh. *J. Food Sci.* 35:478–482.

Buffler, C. R. 1993. *Microwave Cooking and Processing*. New York: AVI, Van Nostrand Reinhold, p. 120.

Dizik, N. S., and Knapp, F. W. 1970. Avocado polyphenoloxidase: Purification and fractionation on Sephadex thin layers. *J. Food Sci.* 35:282–286.

Esaka, M., Suzuki, K., and Kubota, K. 1987. Effects of microwave heating on lipoxygenase and trypsin inhibitor activities and water adsorption of winged bean seeds. *J. Food Sci.* 52:1738–1739.

Halpin, B. E., and Lee, C. Y. 1987. Effect of blanching on enzyme activity and quality changes in green peas. *J. Food Sci.* 52:1002–1005.

Institute of Food Technologists Expert Panel. 1989. Microwave food processing. *Food Technol.* pp. 117–126.

Khalil, H., and Villota, R. 1988. Comparative study on injury and recovery of *Staphylococcus aureus* using microwaves and conventional heating. *J. Food Protect.* 50:181–184.

Lowry, O. H., Rosebrough, N. J., Farr, A. L., and Randal R. J. 1959. Protein measurement with phenol reagent. *Biochem. J.* 193:265–267.

Martinez, M., and Whitaker, J. R. 1995. The biochemistry and control of enzymatic browning. *Trends Food Sci. Technol.* 6:195–200.

Montgomery, M. W., and Petropakis, H. J. 1980. Inactivation of Bartlett pear polyphenoloxidase with heat in the presence of ascorbic acid. *J. Food Sci.* 45:1090–1091.

Muftugil, N. 1986. Effect of different types of blanching on the color, the ascorbic acid and chlorophyll of green beans. *J. Food Proc. Preserv.* 10:69–76.

Owusu-Ansah, Y. J., and Marianchak, M. 1991. Microwave inactivation of myrosinase in canola seeds: A pilot plant study. *J. Food Sci.* 56:1372–1374.

Pedrero, D., and Pangborn, R. 1989. *Evaluación Sensorial de los Alimentos. Métodos Analíticos*, pp. 78–84. Ed. Alhambra Mexicana, Mexico.

Quenzer, N. M., and Burns, E. E. 1981. Effects of microwave steam and water blanching on freeze dried spinach. *J. Food Sci.* 46:410–413, 418.

Rosen, C. G. 1972. Effects of microwaves on food and related materials. *Food Technol.* 26 (7):36–39, 55.

Tajchakavit, S., and Ramaswamy, H. S. 1996. Inactivation kinetics of pectin methylesterase: Thermal vs. microwave heating in batch mode. Evidence of nonthermal effects. *IFT Annual Meeting. Book of Abstracts*, p. 49.

Torringa, H. M., Ponne, C. T., Bartels, P. V., van Remmen, H. H. J., and van Nielen, L. 1992. Microwaves in processing of vegetable produce. *IFTEC Meeting. Book of Abstracts*. The Hague, Netherlands.

Whitaker, J. R. 1994. *Principles of Enzymology for the Food Sciences.* pp. 546–554. New York: Marcel Dekker.

Wrolstad, R. E., Lee, D. D., and Poe, M. S. 1980. Effect of microwave blanching on the color and composition of strawberry concentrate. *J. Food Sci.* 45:1573–1577.

CHAPTER 13

Modeling and Simulating Microbial Survival in Foods Subjected to a Combination of Preservation Methods

MICHA PELEG

INTRODUCTION

T
HE lethal effect of high temperature on microorganisms and spores has been a topic of intensive study during past decades. As a result, there are now generally accepted criteria for judging the safety of existing thermal preservation methods and procedures for calculating the safety of new ones. The impressive safety record of the food industry, especially in its canning and bottling operations, has served as proof of the efficacy of these criteria and procedures. Therefore, although the theory on which these criteria and procedures is based has been continuously challenged and criticized (e.g., Casolari et al., 1988; Casolari, 1994; Cole, 1997), there has been little incentive to revise it. However, the situation has recently changed with the emergence of novel nonthermal preservation methods, such as ultra-high pressure pulsed high-voltage electric fields (Barbosa-Cánovas et al., 1998), and the tendency to reduce the amount of heat in conventional thermal processes in order to preserve freshness and nutrients. There is also a growing interest in new combinations of two or more preservation methods as a compromise between the conflicting demands of ensuring safety and maintaining freshness. Such combinations allow reduction of the overall level of delivered energy through the exploitation of synergistic effects. (In the case of ionizing radiation, as in pork products, this is necessary so that they remain edible). The safety of any new or modified preservation method in a particular product will always require confirmation by direct microbiological analysis. Nevertheless, commercial application of these new technologies will also raise theoretical questions whose answers may be of more than academic interest. The main issues will

163

be how the microbial survival data of novel processes can be used to determine safety and what constitutes an equivalent treatment in processes where the microbial mortality pattern is different. Addressing these areas will obviously require knowledge of the biology, physiology, biochemistry, and biophysics of microorganisms in order to elucidate how the lethal agent works at the molecular and cellular levels. It will also require familiarity with the properties of statistical distributions in order to understand the microbial response to a preservation treatment at the population level. The objectives of this review are to present a statistical approach to the interpretation of various common types of microbial survival curves and to explore the possibility of expressing their characteristics, not in terms of reactions kinetics, but as cumulative forms of distributions of lethal events having different mode, mean, variance, and skewness coefficients.

MICROBIAL SURVIVAL CURVES

The effect of a lethal agent or environment on microorganisms is most commonly expressed in terms of first-order kinetics, such as

$$- dN/dt = kN \tag{1}$$

where N is the momentary number of organisms, t the exposure time, and k a rate constant having time^{-1} units. Integration of Equation (1) gives the familiar relationship

$$N = N_0 \exp [- (kt)] \tag{2}$$

where N_o is the initial number of organisms or

$$\log S = - kt \tag{3}$$

where S is the survival ratio defined as $S = N/N_o$.

Because experimental plots of log S vs. t are frequently considered linear or approximately linear, it has been commonly assumed that microbial mortality, particularly as a result of heat treatment, indeed obeys first-order kinetics. Hence, non-linear log S vs. t relationships, whenever encountered, have usually been considered as evidence that microbial populations were a mixture of two or more sub-populations, each having first-order mortality kinetics but with a different rate constant k (Stumbo, 1973). Or alternatively, they have been considered as evidence of higher order kinetics that could be described by a variety of mathematical models (e.g., those by Rodriguez et al., 1992; Whiting, 1995; Hills and Mackey, 1995; Holdsworth, 1997). The deficiencies of the

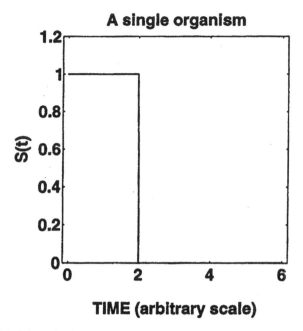

Figure 13.1 Schematic view of the survival curve of a single microorganism or spore.

first-order kinetics approach have been exposed by several investigators, most recently by Peleg and Cole (1998), so there is no need to repeat them. Suffice it to say that an alternative view to the concept of first or higher order kinetics is to consider the survival curve as a cumulative form of a temporal distribution of lethal events. The distribution is at least a partial reflection of the microbial population's spectrum of resistances or sensitivities to the lethal agent. A high temperature, as in the case of a heat treatment, can serve as an example (Linton et al., 1995; Anderson et al., 1996; Augustin et al., 1998). According to this approach, an individual organism, *i,* after an exposure of time, *t,* is either alive or dead (Figure 13.1), or, in the case of bacterial spores, is either viable or not. Hence, if nonfatal injuries and adaptation are excluded, the survival curve of the individual organism or spore is given by (Peleg, 1996a; Peleg et al., 1997)

$$\text{For } t < t_{ci} \qquad Si = 1 \text{ (alive)} \qquad (4)$$

$$\text{For } t \geq t_{ci} \qquad Si = 1 \text{ (dead)}$$

where t_{ci} is the time when the organism dies. For the sake of simplicity, the term organism in what follows will refer to both vegetative cells and spores. Had all the organisms in the population had the same t_{ci}, then the survival curve of the population would also have the shape of the step function (Figure 13.1) that Equation (4) describes. In reality, it is more likely that there is a certain degree of variability in the population (and/or in the efficiency of the delivery mechanism) and, hence, the spectrum of the t_{ci}. In a discrete population, the survival curve would then be given by

$$S = \Sigma \Delta \phi_i t_{ci} \tag{5}$$

where $\Delta \phi_i$ is the fraction of organisms having the same t_{ci} ($\Sigma \Delta \phi = 1$).

In most cases of microbial populations, the t_{ci} spectrum, especially in raw foods, is very large and can be treated as continuous. For large populations, Equation (5) can thus be written in the integral form of

$$S = {}_0\int^1 f\,[t,t_c(\phi)]d\phi \tag{6}$$

where f is a function of time and the t_{ci} distribution.

One of the most flexible and convenient functions to describe survival curves is the Weibull distribution (also known as the Rosin-Ramler distribution in certain engineering fields). Its frequency (or density) form is

$$dS/dt = nbt^{n-1}\exp(-bt^n) \tag{7}$$

and its cumulative form is

$$S = \exp{-bt^n} \tag{8}$$

or

$$\log_e S = -bt^n \tag{9}$$

Because in microbiology base ten is more common, Equation (9) can also be written as

$$\log_{10} S - bt^n/\log_e 10 \tag{9a}$$

According to this model, the log S vs. t plot has a downward concavity whenever $n > 1$ and an upward concavity whenever $n < 1$. As shown in Figure 13.2, a linear relationship among semi-logarithmic coordinates is the basis of the first-order kinetics interpretation and can be considered a special case of the Weibull distribution where $n = 1$. There is ample evidence that many actual survival curves indeed have an upward or downward concavity depending on the organism and the kind of lethal agent (Troller and Christian, 1978; Hills and Mackey, 1995; Holdsworth, 1997; Peleg and Cole, 1998). When survival

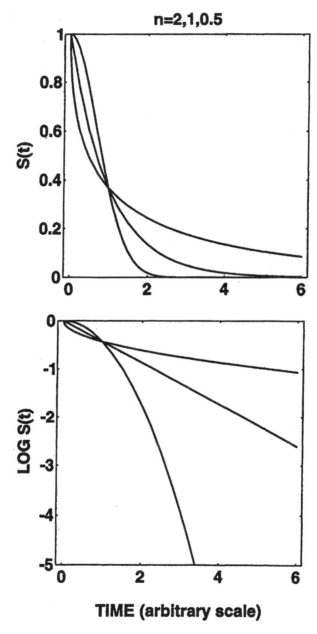

Figure 13.2 Simulated microbial survival curves using the commutative Weibull distribution function on linear and semi-logarithmic coordinates. Note that what appears as first-order mortality kinetics can also be described by a Weibull distribution function with $n = 1$.

curves are determined under different conditions, after heat treatment at different pH levels (Ama et al., 1994), or after exposure to ultra high pressure at different temperatures (Ballestra et al., 1998), there is usually not only a shift in the survival curves' location along the time axis, but also a noticeable change in their concavity. According to the traditional concept, a concave survival curve when plotted on semi-logarithmic coordinates is indicative of a mixed microbial population, each obeying a first-order mortality kinetics or mortality kinetics of a higher or more complex order. Adhering to these explanations requires one to conclude that the application of a combination of preservation methods or merely a change in pH or temperature creates a new microbial mixture or mortality kinetics order. According to the concept presented here, the spectrum of resistances of the same population is different when the conditions are different. Or in other words, a change in pH or temperature is manifested in the distribution properties and not the mortality kinetics. It is acknowledged, of course, that the mechanisms that cause mortality can be or are affected by changes in the conditions, but the focus here is only on their manifestation at the population level.

SIMULATION OF THE EFFECT OF COMBINED TREATMENTS

Let us assume that the population in question is not a mixture and has a unimodal distribution of resistances to the lethal agent. Mixtures can also have a unimodal distribution (Everitt and Hand, 1981), and this will be discussed separately below. The most common parameters that describe the properties of unimodel distributions are the mode (peak location), mean (first moment), variance (a spread measure), and coefficient of skewness. A coefficient of skewness bigger than one indicates a skew to the right, and smaller than one indicates a skew to the left. A symmetric distribution has a coefficient of skewness equal to one. In the case of the above-mentioned Weibull distribution, the mode t_{cm} mean $\bar{t}_c$, variance, σ^2_{tc}, and coefficient of skew ν_1, are given by (Patel et al., 1976):

$$t_{cm} = [(n-1)/nb]^{1/n} \tag{10}$$

$$t_c = \{\Gamma[(n+1)/n]\}/b^{1/n} \tag{11}$$

$$\sigma_{tc}^2 = \{\Gamma[(n+2)/n]-\Gamma[(n+1)/n])^2\}/b^{2/n} \tag{12}$$

and

$$\nu_1 = \mu_3/\mu_2^{3/2} \tag{13}$$

where Γ is the gamma function and $\mu_3 = \Gamma(1+2/n)/b^{3/2}$ and $\mu_2 = \Gamma(1+2/n)/b^{2/n}$. Similar expressions can be found for other distributions (Patel et al., 1976).

The properties of the Weibull distribution parameters are expressed in terms of b and n, the constants of Equation (7).

Let us imagine that by introducing a combined treatment (e.g., as in lowering the pH in a heat treatment), a microbial population becomes more sensitive and its spectrum of resistances not only shifts to lower values as a whole, but also becomes narrower. The effect can easily be visualized (see Figure 13.3) and expressed in terms such as a lower mode and smaller variance. The relationship between the distribution's mode and variance and the introduced change (temperature, pH, number of electric pulses, and pressure) can also be expressed mathematically with a variety of empirical or semi-empirical (e.g., exponential) models. The problem in developing such models is that the mathematical relationship between the Weibull distribution mode and variance (or other parameters) and the constants b and n is not intuitively clear. Also, while the calculation of t_{cm}, $\bar{t}_c$, σ^2_{tc}, or v_1 from the known values of b and n is a trivial matter, the inverse (i.e., calculating b and n from known values of the mode, variance, or a pair of any other parameters) is not. It can be done, though, by numerical methods that are now widely available as part of standard commercial mathematical software. The simulations shown in Figures 13.3 through 13.5 were, in fact, made using the Mathematica® program whose notation will also be used here. Thus, for any given pair of mode and variance values (tm and V), the corresponding values of n and b are calculated from (Peleg, 1996b):

$$n = n/.FindRoot [v / tm\wedge 2== (n/ (n-1))\wedge(2/n)*$$

$$(Gamma [(n+2) / n] - (Gamma [(n+1) / n]\wedge 2)$$

$$, \{n, ng\} \quad] \tag{14}$$

where n_g is an estimated or guessed value of n and

$$b = (n-1) / n / tm\wedge n \tag{15}$$

These conversions are convenient for describing changes in distributions where $n > 1$, and they can be used to generate the curves of the kind shown in Figure 13.3. The direction of the changes depends on the context. With the shift taken from right to left, the figure qualitatively describes the effect of lowering the pH on microbial survival after a heat treatment at a constant temperature, the effect of the temperature increase during an ultra high pressure treatment, or the effect of the pressure increase in a treatment at a constant

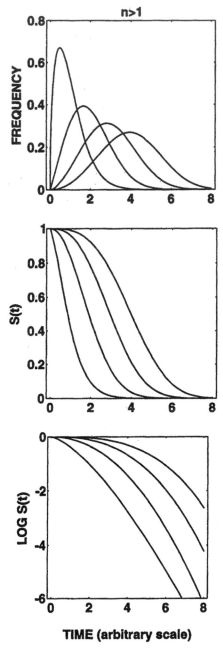

Figure 13.3 Simulated sequence of the spectra of microbial resistances to a lethal agent and the corresponding survival curves plotted on linear and semi-logarithmic coordinates. Note the downward concavity of the semi-logarithmic survival curves.

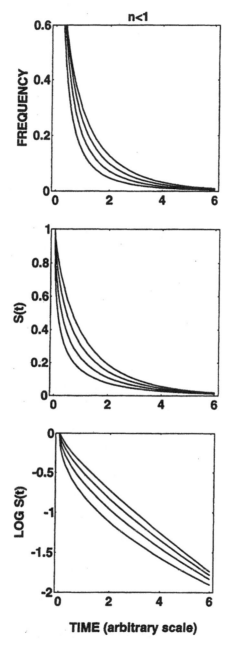

Figure 13.4 Simulated sequence of the spectra of microbial resistances to a lethal agent and the corresponding survival curves plotted on linear and semi-logarithmic coordinates. Note the upward concavity of the semi-logarithmic survival curves.

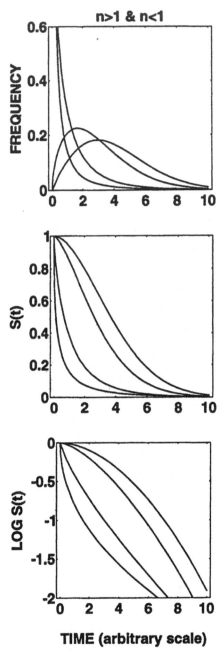

Figure 13.5 Simulated sequence of the spectra of microbial resistances to a lethal agent and the corresponding survival curves plotted on linear and semi-logarithmic coordinates. Note the shapes of the frequency distributions and the concavity inversion in the semi-logarithmic survival curves.

temperature (Peleg and Cole, 1998). In all these cases, the added factor lowers the distribution's mode and reduces its variance. Obviously, Equations (14) and (15) cannot be used for Weibull distributions when $n < 1$ where the mode does not exist. In such a case, n and b can be calculated from the known or assumed mean, mt, and variance (V). For example:

$$n = n/.FindRoot [v= =$$

$$(Gamma[(n+2) / n] - (Gamma [(n+1) / n]^2)*$$

$$mt^2/ (Gamma [(n+1) / n])^2, \{n, ng\}] \qquad (16)$$

and

$$b = (Gamma [(n+1) / n])^n / mt^n \qquad (17)$$

The curves shown in Figure 13.4 were created using these equations. Qualitatively, they can describe changes in the thermal resistance of pathogens such as *Salmonella* or *Listeria* as affected by temperature and/or pH. That these organisms indeed have a concave upward semi-logarithmic survival curves has been reported by several authors (e.g., Ellison et al., 1994; Augustin et al., 1998). Equations (16) and (17) can be used for any type of a Weibull distribution irrespective of whether $n < 1$ or $n > 1$. At least in principle, they can, therefore, be used to simulate process combinations that result in an inversion of the survival curve's concavity. Demonstration of such a simulation is given in Figure 13.5. Although actual cases of concavity inversion as a result of a combined treatment are unknown to the author, their existence cannot be ruled out.

Theoretically, the values of n and b can be calculated numerically using any two distribution parameters. This includes the coefficient of skewness, which together with the mode, mean, or variance can be used for simulations of the kind already shown. Because of its relative mathematical complexity [Equation (13)] and familiarity, it is unlikely that it will be widely used for simulating survival curves. A special case where it could be used profitably is when the combined treatment results in a dramatic change in the distribution's skewness or reversal. An example of such a simulation is shown in Figure 13.6. It ought to be remembered, though, that because of its mathematical structure, the Weibull distribution is not a good model of populations with a notable skew to the left because its coefficient of skewness is always positive. The result can be that choosing Equation (13) as one of the equation parameters to calculate n and b numerically need not yield a unique solution or any solution at all, because for certain values of v_1 there is no converging. If a skew reversal is indeed a prominent characteristic of a combined preservation

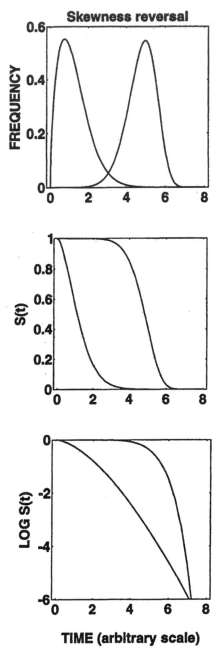

Figure 13.6 Two simulated spectra of microbial resistances to a lethal agent and the corresponding survival curves plotted on linear and semi-logarithmic coordinates. Note the skewness reversal in the frequency distributions and its manifestation in the survival curve shapes.

174

process, it is perhaps better to use another type of a distribution function. One of the most suitable for this particular purpose is probably the beta or log-beta distribution function (Barret et al., 1991). The log-beta frequency (density) distribution has the same structure as the original beta distribution function (Patel et al., 1976):

$$f(x) = x^{p-1}(1-x)^{q-1} /\mathbf{B}\,(p, q) \tag{18}$$

where $\mathbf{B}(p, q)) = \Gamma(p)\Gamma(q)/\Gamma(p + q)$ and $p > 0$, $q > 0$, except that

$$x = \log\,(t/t_{\min})/\log\,(t_{\max}/t_{\min}) \tag{19}$$

where x is the normalized variable, $0 \leq x \leq 1$, t the independent variable, and $t_{\min}$ and $t_{\max}$ are the lower and upper limits of the distribution's domain. Notice that applying the log-beta distribution to microbial survival data entails a finite time at which there will be absolutely no survivors. Whether such a situation indeed exists is a debatable issue outside the scope of this chapter. In any case, $t_{\max}$ can be selected in such a way as to cover any conceivable practical range.

THE THERMAL DEATH TIME

According to Stumbo (1973), a thermal death time in heat preservations τ is the time required to reduce the microbial or spore population by 12 orders of magnitude. Or in terms of survival, τ is the time that corresponds to $S = 10^{-12}$. Experimental survival curves rarely cover more than 5 to 7 decades of microbial load reduction. Therefore, how reasonable the criterion of $S_r = 10^{-12}$ is remains a controversial issue. Let us assume that commercial sterility is defined by $S_r = 10^{-d}$, d being 12 or any other number agreed upon on the basis of experience or theoretical safety considerations. From a purely formalistic viewpoint, any process or process combination that produces the same S level should thus be considered equivalent. Consequently, if the survival curve over the whole pertinent time range can be represented by the Weibull distribution, then the thermal death time τ of any treatment can be given by

$$\tau = (d/b)^{1/n} \tag{20}$$

where b and n are the constants of the corresponding Weibull distribution. [Obviously if $S = 10^{-12}$ is maintained as the mortality criterion, then the τ of any thermal or nonthermal process will be $(12/b)^{1/12}$]. The obvious danger with this formal equivalency is that it does not take into account what survivors (if any) can do in different environments. Therefore, the same initial microbial destruction level caused by different processes need not be an absolute guarantee of product safety unless the survival level is so low that recovery becomes impossible under any circumstances.

MIXED POPULATIONS

There can be situations where the population is truly a mixture of resistant and sensitive organisms. However, this should be established independently by analyzing the survival pattern of organism grown after partial treatments. Only in extreme cases when the distribution of resistances is clearly bi- or multimodal will the existence of a mixture be evident in the shape of the survival curve. An example of such a case is shown in Figure 13.7. In general, mathematical characterization of a bimodal distribution requires at least five parameters: two modes or means, two standard deviations, and the fraction of the two populations (w and $1 - w$). Or in our case:

$$S(t) = wS_1(t) + (1-w)S_2(t) \qquad (21)$$

where $S_1(t)$ and $S_2(t)$ are the survival curves of the two sub-populations.

In the case of a multimodal distribution, the survival curve will be given by

$$S(t) = \Sigma w_i S_i(t) \qquad (22)$$

Dealing with such situations is difficult not only because of the number of constants that are involved, but also because the resistance distributions of the sub-populations need not be all of the same type. Consequently, although Eq (22) can easily be used to simulate the mortality patterns of mixed microbial populations, it may not be a practical model to analyze experimental survival data. This is particularly the case when, as a result of a combined treatment, the response of each sub-population can change in a different manner. A possibility also exists that a mixed population has a unimodal distribution of resistances, as shown in Figure 13.8. In such a case, and especially when the data have a "noise," the curve can be mistaken for that of a single organism if judged solely by statistical criteria (Peleg et al., 1991). Therefore, it is worth repeating that when a mixture is suspected, its existence should be confirmed or denied independent of tests and not on the basis of the survival curve's shape.

CONCLUDING REMARKS

The concept presented here is that a microbial population has a different spectrum of resistances to a lethal agent under different conditions. This will

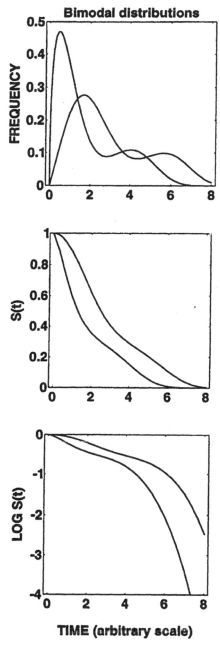

Figure 13.7 Simulated spectra of microbial mixtures to a lethal agent and the corresponding survival curves plotted on linear and semi-logarithmic coordinates. Note the bimodal distributions and their manifestation in the survival curve shapes.

177

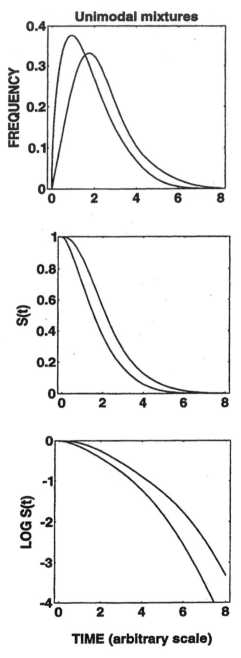

Figure 13.8 An example of a simulated true microbial mixture having a unimodal distribution of resistances and its corresponding survival curves. Note that the latter are practically indistinguishable from those of single organisms.

178

be manifested in the shape of the survival curve, which is the cumulative form of the resistance distribution. In combined methods of preservation, the aim is to increase the microbial mortality rate by lowering the organism's resistance to the treatment. However, because even a unimodel distribution is characterized by at least two parameters (the mode or mean and the variance being the most commonly specified), the combined process can have a different effect on each parameter. Thus, just lowering the microbial population's mode or mean resistance may be insufficient to secure safety if the distribution is wide enough to allow for the survival of a sufficient number of resistant organisms. However, when two or more combined treatments are truly synergistic, they will almost certainly lower the resistance distribution mode and simultaneously reduce its spread (Peleg and Cole, 1998). When the semilogarithmic survival curve of the target organism is concave upward, the vast majority of the population is destroyed within a very short time. But a very small number of resistant individuals can still survive, as indicated by a highly skewed resistance distribution with a long tail to the right. Microbial safety in this case would be determined not so much by the actual number of organisms killed, but by the ability of the survivors to restore their populations in the processed food.

The single model used in this work was based on the Weibull distribution function. As shown, it was flexible enough to account for different kinds of mortality patterns that in the past required for their explanation the assumption of different kinetic orders or the appearance of a new mixture whenever conditions changed. With this model, the most commonly encountered mortality patterns can be attributed to differences in a resistance distribution mode, mean, variance, and skewness coefficient. Or once transformed into the Weibull distribution equation, they could be described in terms of its constants b and n. The Weibull distribution was primarily selected because of its mathematical convenience. Other distributions (e.g., the log-normal, log-logistic, log-beta functions) could probably be just as appropriate or even more suitable in certain situations. In fact, it can be shown that the log-beta function would be clearly more suitable for any process combination that results in a reversal of the population's resistance distribution skewness. It can also be shown that experimental data with a clear unimodel distribution (microbial survival data included) can usually be described by several functions with the same degree of fit. The choice of the distribution function will hardly make a difference if only the regions at the center of the domain were of concern. It can be a totally different matter if the distribution's tail is of practical significance. In fact, different models can provide qualitatively different estimates of the ultimate survival ratio, zero vs. finite. Consequently, the number of survivors estimated by a mathematical model would be extremely sensitive to the properties of the model selected. Hence, it would not be advisable to use any mathematical model for extrapolation unless there are compelling reasons to do so based on physical or biological mechanisms.

ACKNOWLEDGEMENT

The author wishes to thank the Massachusetts Agricultural Experiment Station at Amherst for their contribution.

REFERENCES

Ama, A. A., Hamdy, M. K., and Toledo, R. T. 1994. Effect of heating, pH and thermo radiation on inactivation of *V. vulnificus*. *Food Microbiol.* 11:215.

Anderson, W. F., McClure, P. J., Baird-Parker, A. C., and Cole, M. B. 1996. The application of a log-logistic model to describe the thermal inactivation of *C. botulinum* 213B at temperatures below 121.1°C. *J. Appl. Bacteriol.* 80:283.

Augustin, J. C., Carlier, V., and Roziier, J. 1998. Mathematical modeling of the heat resistance of *L. monocytogenes*. *Appl. Microbiol.* 84:185.

Ballestra, P., daSilva, A., and Roziier, J. 1998. Inactivation of *E. coli* by carbon dioxide under pressure. *J. Food Sci.* 61:829.

Barbosa-Cánovas, G. V., Pothakamury, U. R., Palou, E., and Swanson, B. G., eds. 1998. *Nonthermal Preservation of Foods*. New York: Marcel Dekker, Inc.,

Barret, A. M., Normand, M. D., and Peleg, M. 1991. A "log-beta" vs. the log-normal distribution for particle populations with a wide finite size range. *Powder Technol.* 66:195.

Casolari, A. 1988. Microbial death. In *Physiological Models in Microbiology*, Vol. II. Bazin, M. J., and Prosser, J. I. eds. Boca Raton, FL: CRC Press, pp. 1–44.

Casolari, A. 1994. About basic parameters of food sterilization technology. *Food Microbiol.* 11:75.

Cole, M. B. 1997. The outlook for novel preservation technologies: A food industry prospective. Paper presented at the *COFE 97, AICHE* annual meeting at Los Angeles, November.

Ellison, A., Anderson, W., Cole, M. B., and Stewart, G. S. A. B. 1994. Modeling the thermal inactivation of *Salmonella typhimurium* using bioluminescence data. *International J. Food Microbiol.* 23:467.

Everitt, B. S., and Hand, D. J. 1981. *Finite Mixture Distributions*. London: Chapman and Hall.

Hills, B. P., and Mackey, B. M. 1995. Multi-compartment kinetic models for injury, resuscitation induced log and growth in bacterial cell populations. *Food Microbiol.* 12:335.

Holdsworth, S. D. 1997 *Thermal Processing of Packaged Foods*. London: Blackie Academic and Professional.

Linton, R. H., Carter, W. H., Pierson, M. D., and Hackney, C. R. 1995. Use of a modified Gompertz equation to model nonlinear survival curves for *L. monocytogenes*. Scott, A. *J. Food Protect.* 58:946.

Patel, J. K., Kapadie, D. H., and Owen, D. B. 1976. *Handbook of Statistical Distributions*. New York: Marcel Dekker.

Peleg, M. 1996a. Evaluation of the Fermi equation as a model of dose response curves. *Appl. Microbiol. Biotechnol.* 46:303.

Peleg, M. 1996b. Determinations of the parameters of the Rosin-Ramler and beta distributions from their mode and variance using equation-solving software. *Powder Technol.* 87:181.

Peleg, M., and Cole, M. B. 1998. Reinterpretations of microbial survival curves. *CRC Crit. Rev. Food Sci. and Nutrition* 38(5):353.

Peleg, M., Normand, M. D., and Damrau, E. 1997. Mathematical interpretation of dose-response curves. *Bull. Math. Biol.* 59:747.

Peleg, M., Normand, M. D., and Nussinovitch, A. 1991. On theoretical mixed particle populations with a unimodal size distribution. *Powder Technol.* 68:281.

Rodriguez, A. C., Smerage, G. H., Teixeira, A. A., Lindsay, J. A., and Busta, F. F. 1992. Population model of bacterial spores for validation of dynamic thermal processes. *J. Food Proc. Eng.* 15:1.

Stumbo, C. R. 1973. *Thermobacteriology in Food Processing,* 2nd ed. New York: Academic Press.

Troller, J. A., and Christian, J. H. B. 1978. *Water Activity and Food.* New York: Academic Press.

Whiting, R. C. 1995. Microbial modeling in foods. *CRC Crit. Rev. Food Sci. Nut.* 35:467.

CHAPTER 14

Textural and Structural Changes in Apples Processed by Combined Methods

ADELMO MONSALVE-GONZÁLEZ
GUSTAVO V. BARBOSA-CÁNOVAS
BARRY G. SWANSON
RALPH P. CAVALIERI

INTRODUCTION

THE market for minimally processed fruits and vegetables has grown steadily for the past decade as a result of an increasing consumer demand for fresh-like foods. Minimally processed food, as the term implies, emphasizes a process or technology that leads to the preservation of foods in which organoleptic characteristics, such as texture, flavor, and color, are similar to the fresh counterpart without compromising its wholesomeness (microbial safety of the food).

Combined methods, or hurdle effects, are preservation techniques that could be considered for minimally processed foods. No major modifications of the organoleptic properties of a food are introduced by the use of the techniques, and microbial stability is accomplished (Levi et al., 1985; Gould and Jones, 1989; Leistner, 1992).

Combined methods are preservation techniques that are gaining importance in the preparation of shelf-stable food products, such as fruits and vegetables, that are shelf stable at room temperature. The benefit of combined methods relies on the capability of producing fresh-like products with technological simplicity. Thus, combined methods have become an attractive option to preserve fruits and vegetables in developing countries (Aguilera et al., 1993).

Combined methods restrain microbial growth by applying a combination of stress factors, or hurdles, such as low pH, reduction in water activity (a_w), the presence of antimicrobial agents, and, in some cases, mild heat treatment (Leistner, 1992). The synergistic and/or additive effect of combined stress factors permits the preservation/stabilization of food with improved quality by avoiding the more damaging use of any single preservation technique, such as heat treatment, dehydration, or freezing (Chirife and Favetto, 1992; Leistner, 1992; Gould and Jones, 1989).

One characteristic of the combined method technique is the incorporation of solutes (generally salts or sugars) to decrease the a_w of a food in conjunction with a reduction of pH with organic acids. This, in addition to the use of antimicrobial, bacteriostatic, or fungistatic agents, helps to inhibit the growth of microorganisms (Karel, 1975; Bolin et al., 1983; Leistner, 1985; Levi et al., 1985; Alzamora et al., 1989; Cerruti et al., 1990; Rojas et al., 1991; Leistner, 1992).

Combined methods could be considered an extension of the intermediate moisture food (IMF) concept (Chirife and Favetto, 1992). Nevertheless, particular differences between both processes exist. For instance, combined method products have greater a_w (0.90–0.97) than intermediate moisture food (IMF) products (0.60–0.85); therefore, the need to decrease a_w beyond 0.85 to inhibit the growth of *Staphylococcus aureus* in IMF products requires the use of a high proportion of solutes, and, as a result, physical properties of the preserved food can be negatively modified (Chirife and Favetto, 1992). Foods preserved by combined methods, on the contrary, are designed with an osmotic treatment to reduce a_w in the 0.90–0.97 range where physical properties (e.g., texture) are less affected (Alzamora et al., 1989; Argaiz et al., 1991; Chirife and Favetto, 1992; Monsalve-González et al., 1993a).

Because the osmotic adjustment of a_w is a common step in processing fruits and vegetables by combined methods, much of the research concentrates on improvement of final food quality (Dixon and Jen, 1977; Levi et al., 1983; Levi et al., 1985; Le Maguer, 1988; Alzamora et al., 1989), as well as characterizing the mass transfer phenomena and its relationship to sugar, the sugar-fruit ratio, temperature, and fruit geometry (Farkas and Lazar, 1969; Moy et al., 1978; Conway et al., 1983; Lerici et al., 1985; Marcotte, 1988; Raoult et al., 1989). One aspect of the osmotic adjustment of a_w that has been overlooked is its effect on textural changes (principally softening) and structural modification.

Most textural and structural studies with apples have been conducted with fresh tissues (Ben-Arie and Kislev, 1979; Abbott et al., 1984; Kovács et al., 1988; Abbott et al., 1989; Glenn and Poovaiah, 1990). Softening in fresh apples is an enzymatic-mediated event that results in the hydrolysis of pectin substance in the middle lamellae and the cell wall, with a parallel increase in soluble pectin (Poovaiah, 1986; Glenn and Poovaiah, 1990). However, softening of the apple tissue during osmotic treatment and adjustment of a_w may be dependent on physical and chemical changes and perhaps less dependent on enzymatic activity. In that regard, Kanujoso and Luh (1967) hypothesized that the transformation of protopectin to water-soluble pectin is involved in the softening of canned peaches. Forni et al. (1986) reported a significant correlation between the amount of protopectin that was leached from two apricot cultivars and softening during osmotic dehydration. Additionally, diffusion of sucrose to the intercellular spaces (Levi et al., 1983), loss of turgor, and ion movement from the cell wall (Poovaiah, 1986) may be other operating mechanisms of softening.

Calcium is used to decrease undesirable softening in fresh apples during storage. Calcium is believed to act by two general mechanisms. One involves regulation of cell function, such as changes in cell wall structure and membrane permeability. The other is the interaction of calcium with pectin to form a cross-linked polymer network that increases mechanical strength (Poovaiah, 1986; Glenn and Poovaiah, 1990). Infiltration of calcium into fresh apple tissue and processed apples has been utilized to maintain firmness (Mason, 1976; Johnson, 1979; Sams and Conway, 1984; Abbott et al., 1989; Glenn and Poovaiah, 1990). From ultrastructural studies, infiltrated calcium is shown to bind to cell walls and the middle lamellae where it exerts its major influence on firmness (Glenn and Poovaiah, 1990). The objectives of this study were to relate textural and structural changes induced by the osmotic adjustment of a_w and to examine the role of calcium in preventing the softening of apples preserved by combined methods.

MATERIALS AND METHODS

FRUIT SAMPLES

Red Delicious (*Malus domestica,* Borkh) apples were obtained from a commercial orchard in Yakima, Washington. Apples were transported to Washington State University (WSU) and were placed in controlled atmospheric storage (1% O_2, 1% CO_2, 95% humidity, and 1°C) until further use. The controlled atmospheric storage of the fresh apples did not exceed three months. For experimental purposes, apples of similar firmness (63 to 67 Newtons) and solid content (12 to 14° Brix) were chosen for preservation by combined methods.

ADJUSTMENT OF WATER ACTIVITY

Medium- to large-sized (125 to 235 g) apples were simultaneously peeled, cored, and sliced by a mechanical apple peeler (White Mountain, model 300). The peeler-corer device yielded 1-cm thick apple slices. Cylindrical sections were produced by a perpendicular cut through the center-most area of the apple slices with a cork borer (1.1 cm diameter). Slices and cylinders were placed in glass beakers with an acidified and a non-acidified sucrose solution (52° Brix). The sucrose solutions contained 4-hexyl resorcinol as an anti-browning agent (OPTA Food Ingredients, Inc., Cambridge, MA) and were added to the fruit in a 4:1 ratio (w/w). The solution concentrations were selected to reduce the apple a_w to the equilibrium range of 0.94 to 0.96.

Ascorbic acid (0.21%) and 4-hexyl resorcinol (0.0021%) were added to preserve the color of the product during the osmotic adjustment of a_w (Monsalve-González et al., 1993b). The beakers containing apple sections were transferred to a constant temperature orbital water bath shaker (Magni Whirl, GS Blue M Electric, Blue Island, IL). The suspensions were maintained at the chosen temperatures (30, 40, and 50°C) and were agitated at 150 rpm for 15 min to 12 hours.

Processed apple slices or cylinders were withdrawn from the beakers at regular intervals and were immediately blotted with absorbent tissue to remove excess solution. Processed fruits were allowed to equilibrate to room temperature in closed, small glass containers for 12 hours before being analyzed for moisture, sugar uptake, a_w, and texture.

WATER ACTIVITY MEASUREMENT

Apple slices and cylinders were allowed to equilibrate in closed glass containers for 12 hours at room temperature ($22 \pm 1°C$). A small void volume was left in the container to speed up the moisture redistribution among the apple slices and cylinders. The a_w was measured using an Aqua Lab CX-2 electrical hygrometer (Decagon Devices, Inc., Pullman, WA). Unless otherwise stated, a_w determinations were conducted in triplicate, and the mean was reported.

TEXTURAL CHANGES AND CALCIUM TREATMENT

Calcium chloride ($CaCl_2 \cdot 2H_2O$) was added to a new set of acidified and non-acidified sucrose solutions (52° Brix) to make up 0.15%, 0.30%, 0.60%, and 1% (w/w) concentrations. Apple cylinders were treated with the selected solutions at 30°C to adjust the a_w. Apple cylinders were removed from the solutions at regular intervals and were immediately blotted with absorbent tissue to eliminate excess sucrose solution. Compression forces (F_c) of the apple cylinders were determined between two flat plates with an Instron Testing Machine Model 1350 (Instron Corporation, Canton, MA). Fifteen fresh and treated apple cylinders from the same apple were compressed between the two flat plates to 25% of their original height (7.5–6.5 mm) beyond rupture of the tissue. Because texture is not uniform among fruits, a force ratio between the treated and fresh apple cylinders was established to study the dependence on treatment time and temperature. The speed of the plate was set at 1 mm/sec, and the average determination with the 95% confidence level was reported.

SCANNING ELECTRON MICROSCOPY AND ENERGY DISPERSIVE X-RAY ANALYSIS

Apples with similar firmness were selected to carry out the structural studies. Only two slices obtained from the center-most area of the fruit and perpendicular to the stem/blossom axis were assayed. The slices were osmotically treated with acidified sucrose solution (52° Brix) at 30°C. Treated apple slices used for microstructural and dispersive X-ray studies were prepared in triplicate. A replicate consisted of two slices per apple with the a_w adjusted to 0.94 to 0.96 using independent osmotic treatments. Comparisons between fresh and treated apple slices were made using slices obtained from the same apple to minimize variability.

Treated and fresh apple slices were cut in small cubes (5 × 5 mm) using double-sided razor blades. Perpendicular cuts were performed through the middle portion of the slice between the stem-calyx axis and the peel. Apple cubes were marked by making a 45° cut at the corner of the specimen to identify the orientation of the cube toward the skin and core.

Apple cubes were frozen in liquid nitrogen and freeze-dried in a Virtis freeze drier (The Virtis Company, Inc., Gardiner, NY) at −30°C for four days. Immediately after processing, apple cubes were fixed overnight with 3.0% glutaraldehyde in a 0.1 M Pipe buffer (pH 7.2) at room temperature. Following fixation, the apple pieces were rinsed with two changes of fresh 0.1 M Pipe buffer for 10 min each to remove excess glutaraldehyde. Thereafter, the fixed apple pieces were dehydrated in a graded ethanol series from 10% to 100% as follows: 10% increments from 10 to 70% for 10 min each; 80 to 95% for 10 min each; and three changes at 100% ethanol for 15 min each. Dehydrated apple pieces were critical-point dried in a Bomar SPC 1500 using carbon dioxide as the transitional fluid (Trakoontivakorn et al., 1988). The freeze- and critical-point dried sections were mounted on aluminum specimen mounts with double-sided adhesive tape. Sections were carbon coated in an ETEC Evaporator and were gold coated in a Hummer V Sputtering apparatus (Hummer-Technics) with a 30-nm layer of gold.

X-ray counts for calcium were estimated by integrating the peak at the window width of 330 EV. Each specimen of either treated or fresh apple was viewed for 120 sec and in triplicate in a Hitachi S570 Scanning Electron Microscope (SEM) equipped with a KEVEX Micro-X Analytical Spectrometer 7000 and Energy Dispersive X-ray (EDAX) to construct an element map of calcium. The following parameters were kept constant for SEM throughout the analysis: (1) accelerating voltage of 15 kV; (2) tilt 25°; (3) magnification 120 ×; and (4) 13 to 15 mm of working distance.

STATISTICAL ANALYSIS

Analysis of variance and linear regression of the data were analyzed using the statistical Analysis System (SAS Institute, 1985) at the significance level of $\alpha = 0.05$.

RESULTS AND DISCUSSION

TEXTURE CHANGES AND STRUCTURAL CHANGES DURING ADJUSTMENT OF a_w

Within 2 hours of osmotic treatment with sucrose, apple a_w decreased from 0.989 (fresh) to 0.974 (treated). For apple slices immersed in the sucrose solution for 10 hours the average a_w was 0.951 ± 0.002. Texture modification as a result of the osmotic treatment is shown in Figure 14.1(a)–(c). A decrease in F_c, or softening of the apple cylinders, occurred within the first 2 hours of a_w osmotic adjustment [Figure 14.1(a)]. No apparent effect of solution temperature was evident on softness as indicated by a similar tendency in F_c values beyond the 2 hour treatment [Figure 14.1(a)–(c)].

The F_c ratio decreased more by osmotic treatment using an acidified sucrose solution as compared to non-acidified sucrose solution treatments at 40 and 50°C [Figure 14.1(b) and (c)]. However, solubilization of protopectin may account for some of the softening during the osmotic treatment (Kanujoso and Luh, 1967; Forni et al., 1986). Forni et al. (1986) reported a good correlation between the loss of protopectin and texture during the osmotic dehydration of apricots and peaches. The greater loss of protopectin in peaches (about 50%) compared with apricots (about 20%) during processing rendered the treated peaches with a poor texture and organoleptically unacceptable.

A significant relationship was observed between the F_c values and sugar uptake independent of the treatment temperature (Table 14.1). As the sugar uptake increased, the F_c ratio decreased, perhaps implying a relationship between softening, sugar uptake, and water loss (Monsalve-González et al., 1993a). During osmotic treatment, apple pieces take up sucrose and lose water due to differences in the chemical potential between fruit and osmotic solutions until a hypothetical chemical equilibrium is reached. This mass transfer phenomenon may influence the observed softening by several factors, such as (1) calcium leaching from the cell wall; (2) loss of cell turgor; (3) degradation of the middle lamellae by enzyme activity (Poovaiah, 1986); and (4) solubilization of protopectin (Forni et al., 1986). These factors, acting together or independently, may contribute to the loss of crisp texture in fresh apples and the softer texture of osmotically treated apples.

Softening in the processed apples was limited to the edges of the slices and cylinders where the sucrose solution penetrated during the first 2 hours of the a_w adjustment treatment (Monsalve-González et al., 1993a). The F_c ratios of the osmotically treated and fresh apple cylinders were not significantly differ-

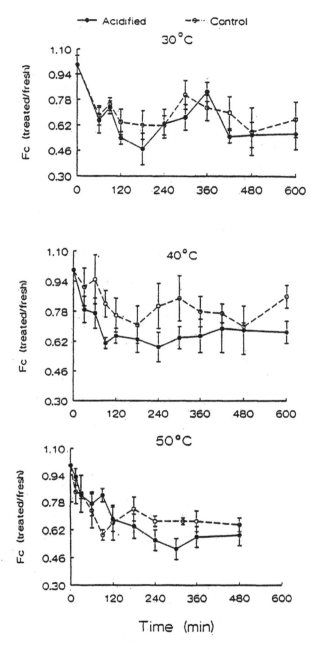

Figure 14.1 Compression force ratio (F_c) changes in apple cylinders during the osmotic adjustment of a_w in an acidified and a non-acidified sucrose solution at different temperatures. Upper figure (a): ratio of hardness (Newton) between treated apples and fresh apples at 30°C. Middle figure (b): 40°C. Lower figure (c): 50°C.

TABLE 14.1. Correlation Coefficients between Sugar Uptake and Compression Force (F_c) as Influenced by Temperature.

Sugar Gain gr./gr. Fresh Apple	Texture Non-Acidified 30°	Texture Acidified 30°C	Texture Non-Acidified 40°C	Texture Acidified 40°C	Texture Non-Acidified 50°C	Texture Acidified 50°C
Acidified 30°C	-0.886*	-0.74*	—	—	—	—
Non-acidified 30°C	-0.864*	-0.65**	—	—	—	—
Acidified 40°C	—	—	-0.68**	-0.77*	—	—
Non-accidified 40°C	—	—	-0.63**	-0.78*	—	—
Acidified 50°C	—	—	—	—	-0.87*	-0.93*
Non-acidified 50°C	—	—	—	—	-0.85*	-0.90*

* At least 1% of significance.
** At least 5% level of significance.

ent when the external edges of the treated apple were removed with a razor blade and compared to fresh apple counterparts of similar thickness (Monsalve-González et al., 1993a). Bolin et al. (1983) reported that the migration of sucrose and high fructose syrup into apple tissues is limited to the periphery by histological slicing procedures, but diffusion of the sugar further into the tissue is dependent on sugar type, temperature, and time.

Trakoontivakorn et al. (1988) observed that the breakdown of cell walls and collapse of cell structure were evident in treated apples [Figures 14.2(b), 14.3 and 14.4(b)], whereas control apples (immersed in distilled water) showed the typical cell integrity and cell arrangement network characteristic of fresh apples [Figures 14.2(a) and 14.4(a)]. A more detailed pattern of collapsed cell walls is presented at higher magnifications in Figures 14.3 and 14.5(b). Comparisons between fresh and osmotically treated apples for 3 hours (52° Brix) are shown in Figures 14.5(a) and 14.5(b), respectively. Cell shrinkage was not apparent in apple sections fixed with glutaldehyde and critical-point dried [Figures 14.4(a)–14.5(b)], but slight shrinkage was noticeable in apple sections prepared by freeze-drying [Figures 14.2(a), 14.2(b) and 14.3]. Domingo and Lluch (1991) also reported that chemical fixations introduced less structural modification than freeze-drying in osmotically dehydrated apples. Glenn and Poovaiah (1990) observed less structural preservation in freeze-dried apple sections than in chemically fixed counterpart apple sections. Chemical fixation is known to produce minimal tissue distortion (Lapsley et al., 1992). However, one advantage of freeze-drying over chemical fixation is that sugar deposition in apple tissue can be visualized [Figures 14.2(b) and 14.3].

Osmotically treated apple slices appeared strikingly different than fresh apple slices. The deposition of sucrose at the intercellular sites and an overall collapse of the cell walls were markedly observed [Figures 14.2(b), 14.4(b) and 14.5(b)]. In addition, the cell walls of fresh apples were cut through, indicating a high tensile strength [Figures 14.4(a) and 14.5(a)], while those of the apples osmotically treated for 3 and 10 hours were collapsed, suggesting less cell-to-cell integrity than fresh apples [Figures 14.4(b), 14.5(b) and 14.6]. Cell collapse and decreased cell-to-cell contact were more visible as the length of the osmotic treatment was increased from 3 to 10 hours [compare Figures 14.4(b), 14.5(b) and 14.6]. The cell walls of the osmotically treated apples also appeared distended [Figures 14.5(b) and 14.6], indicating low cell-to-cell contact. Lapsley et al. (1992) observed poor cell cohesion in apple sections obtained from a soft and mealy apple variety (Rubinette cultivar). Weaknesses at the cell wall sites observed as cell separation reportedly resulted from middle lamella breakdown (Lapsey et al., 1992; Glen and Poovaiah, 1990). Deposition of sugar along the intercellular spaces alters the osmotic balance and may influence softening by affecting cell permeability and, thereby, cell turgor (Bolin et al., 1983; Poovaiah, 1986).

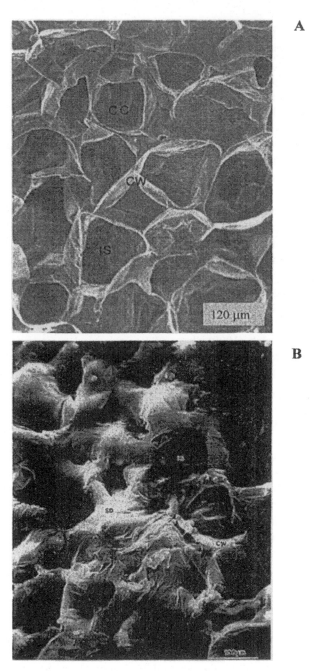

Figure 14.2 Freeze-dried section of a fresh apple slice (a) and osmotically treated apple slice for 3 hours (b) showing sugar deposits (SD), cell wall (CW), and intercellular spaces (IS).

192

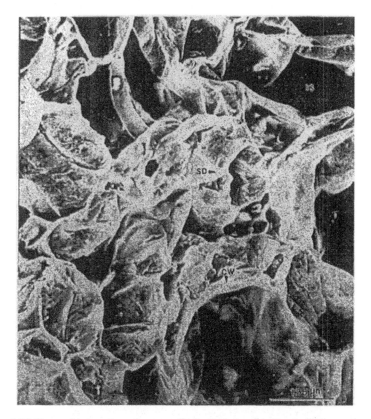

Figure 14.3 Freeze-dried section of an osmotically treated apple slice for 10 hours showing a distorted cell arrangement due to collapsed cell walls and an apparent sucrose deposit (SD).

CALCIUM EFFECT ON SOFTENING

Figures 14.7(a)–(c) illustrate the effect of $CaCl_2 \cdot 2H_2O$ added to osmotic solutions at 0.15% (w/w) and 0.60% (w/w) upon apple softening. A beneficial effect of calcium on the F_c was not observed at the 0.15% concentration. A significant increase ($\alpha = 0.05$) in the F_c was observed as the concentration of calcium was increased to 0.30 and 0.60% [Figures 14.5(b)–(c)]. Concentrations of calcium greater than 0.60% did not improve the overall F_c ratio.

The apparent shape divergence of Figure 14.7(a)–(c) is a reflection of the inherent variability of apple firmness. Each immersion time corresponds to the F_c of treated and untreated apple cylinders obtained from individual apples.

Infiltration of calcium to maintain apple structure is well documented (Poovaiah, 1986; Stow, 1989; Glen and Poovaiah, 1990). Addition of external calcium increases the calcium in the apple tissue and binds to the cell wall

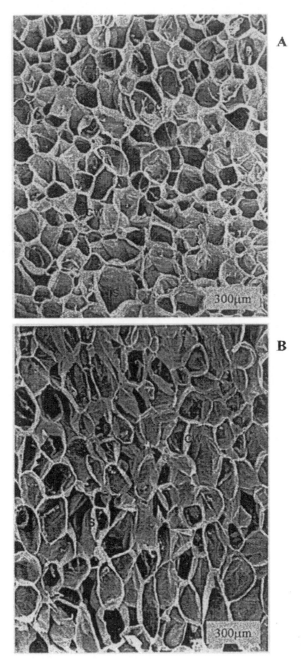

Figure 14.4 Critical-point dried section of a fresh apple slice immersed in distilled water for 3 hours (a) and an osmotically treated apple slice for 3 hours (b) showing cell wall (CW) and intercellular spaces (IS) and cell content (CC).

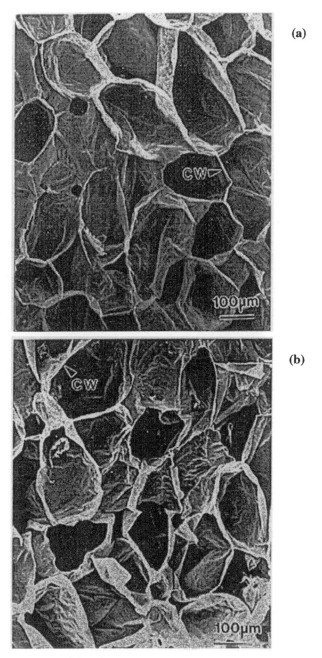

(a)

(b)

Figure 14.5 Critical-point dried section of a fresh apple slice (a) and an osmotically treated apple slice for 3 hours (b) at high magnification showing cell wall (CW), cellular arrangement, and distended cell wall (arrow).

195

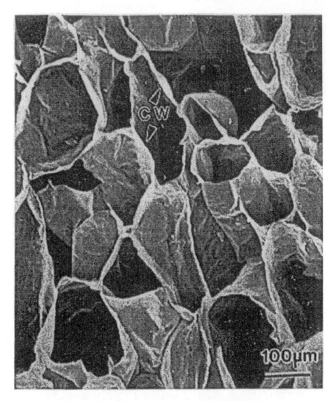

Figure 14.6 Critical-point dried section of an osmotically treated apple slice for 10 hours at high magnification showing a distorted cell arrangement and collapsed cell walls.

to increase the ability of calcium to cross-link pectin and form calcium pectates (Glen and Poovaiah, 1990). Extensive cross-linking of pectin polymers in the cell wall may increase mechanical strength and firmness as illustrated in Figures 14.7(b)–(c).

For the calcium-treated apples, no differences in the F_c ratio were observed between the acidified and non-acidified sucrose solutions. Previous findings suggest that either calcium or protopectin solubilization may be involved in softening.

Apple slices treated with 0.60% (w/w) calcium chloride for 10 hours exhibited minimal cell-wall collapse in comparison to apples osmotically treated with a plain sucrose solution ([Figures 14.8(a)–14.10(b)]). The apple tissue of calcium-treated apples sheared through the cell walls [Figures 14.8(a), 14.9(a), 14.10(a)], indicated high tensile strength, while in the non-Ca-treated apples, the cell walls collapsed due to mealiness and lack of cell cohesiveness [Figures 14.6, 14.8(b), 14.9(b) and 14.10(b)]. This microstructural observation is consistent with the compression test discussed earlier.

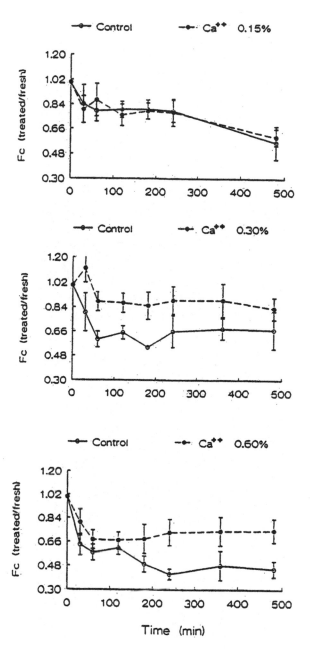

Figure 14.7 Compression force ratio (F_c) changes of an apple cylinder during osmotic adjustment of a_w infiltrated with $CaCl_2 \cdot 2H_2O$ at 30°C. Upper figure (a): apple cylinders treated with 0.15% (w/w). Middle figure (b): 0.30% (w/w). Lower figure (c): 0.60% (w/w).

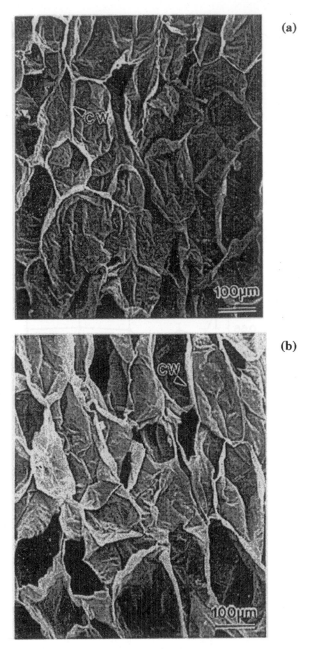

(a)

(b)

Figure 14.8 Critical-point dried section of an osmotically treated apple slice for 3 hours with 0.60% calcium added to the sucrose solution (a) and with no calcium treatment (b) showing cell arrangement. Calcium-treated apple fracture through cells is due to strong cell-to-cell contact, whereas cell walls of non-treated apples are collapsed.

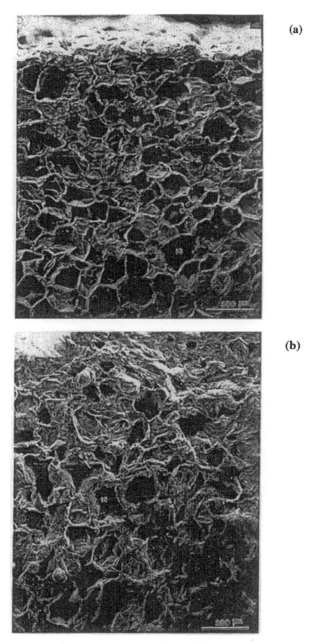

Figure 14.9 Freeze-dried section of an osmotically treated apple slice for 10 hours with 0.60% calcium added to the sucrose solution (a) and with no calcium treatment (b) showing cell arrangement and sucrose deposits (SD). Calcium-treated apple fracture through cells due to strong cell-to-cell contact, whereas cell walls of non-treated apples are collapsed and distorted.

199

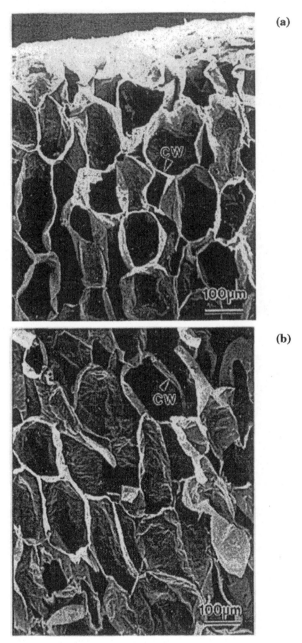

(a)

(b)

Figure 14.10 Critical-point dried section of an osmotically treated apple slice for 10 hours with 0.60% calcium added to the sucrose solution (a) and with no calcium treatment (b) showing cell arrangement. Calcium-treated apple fracture through cells is due to strong cell-to-cell contact, whereas cell walls of non-treated apples are collapsed and distorted.

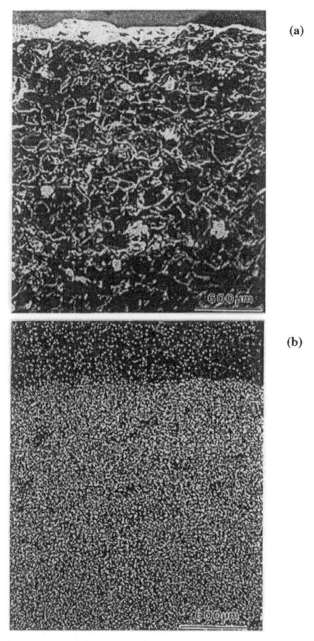

(a)

(b)

Figure 14.11 (a) Freeze-dried section of osmotically treated apple tissue (b) and an element map of the distribution of calcium (white dots).

201

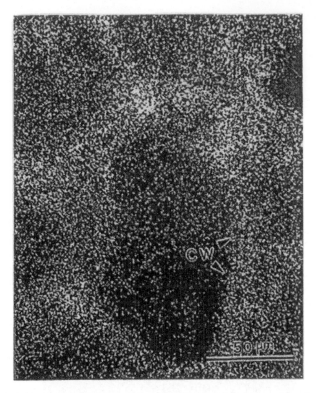

Figure 14.12 Critical-point dried section of osmotically treated apple tissue showing an element map of the distribution of calcium (white dots). Cell wall (CW) is clearly distinguishable in the element map as a dense accumulation of white dots (arrows).

Accumulation of calcium within the tissue of the treated apples as determined by dispersive X-ray analysis was greater toward the edges of the tissue and was smaller toward the center. An element map of calcium distribution in the cell walls is presented in Figure 14.11(b), and a detailed map of calcium is presented in Figure 14.12. The contour of the cell walls is clearly visualized as an accumulation of white dots that correspond to the Ca signal (Figure 14.12) and corroborates the importance of calcium in preserving the structural integrity and, hence, the texture at the cell wall sites. Our results are consistent with reports by Glen and Poovaiah (1990), who claimed that calcium accumulated at the wall of fresh apples infiltrated with calcium chloride is mainly responsible for maintaining cell-to-cell cohesion and apple firmness during storage.

Figures 14.11(a) and (b) present the gradient of calcium (white dots) decreasing in concentration toward the center of the tissue as expected during a diffusion treatment. A plot illustrating the decreasing gradient of the calcium

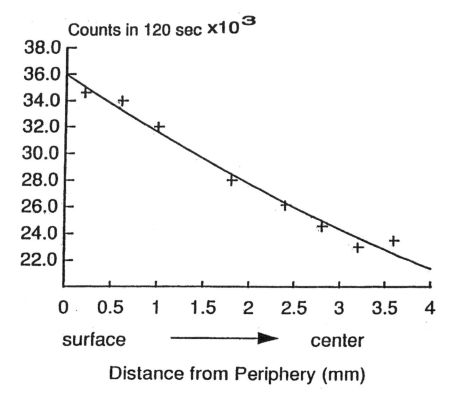

Figure 14.13 Calcium counts by energy dispersive X-ray (120 sec) on apple slices osmotically treated for 10 hours.

distribution in an apple section is presented in Figure 14.13. The number of counts declined as the measurements were taken from the periphery, or external section, of the apple slice in contact with the sucrose solution, to the center of the slice. The exponential trend of Figure 14.13 is concordant with that of the sucrose diffusion into the tissue, which also decreases from the periphery toward the center as determined by assaying the sugar concentration of treated apples (Monsalve-González et al., 1993a).

CONCLUSIONS

After apples were softened by osmotic treatment, a significant negative correlation was observed between sugar diffusion and loss of texture or softening. Softening took place after 2–4 hours of the a_w adjustment treatment. Softening was more pronounced in the acidified sucrose solutions than in the non-acidified sucrose solutions. The correlation between sucrose uptake and texture loss was corroborated by removing the external layers of the apple tissue. As a result, the F_c of the treated apple tissue was similar to fresh apple tissue. Addition of calcium chloride beyond 0.30% (w/w) concentration partially suppressed the softening process. Microstructural and energy-dispersive X-ray studies demonstrated calcium concentrated at the wall sites and helped to preserve the cell structure of osmotically treated apples.

ACKNOWLEDGEMENTS

Funding for this project was provided by the Washington State University (WSU) IMPACT Center. The technical assistance of Dr. Christine Davitt at the Washington State University Electron Microscopy Center is greatly appreciated. This paper is part of the CYTED-D Project.

REFERENCES

Abbott, J. A., Conway, W., and Sams, C. E. 1989. Post-harvest calcium chloride infiltration affects textural attributes of apples. *J. Amer. Soc. Hort. Sci.* 114:932–936.

Abbott, J. A., Wataba, A. E., and Massie, D. V. 1984. Sensory and instrument measurement of apple texture. *J. Amer. Soc. Hort. Sci.* 109:221–228.

Aguilera, J. M., Chirife, J., Parada-Arias, E., and Barbosa-Cánovas, G. V. 1993. CYTED-D AHI: An Ibero-American project on intermediate moisture food and combined methods technology. In *AIChE Symp. Series 297, Food Dehydration.*

Alzamora, S. M., Gerschenson, L. N., Cerruti, P., and Rojas, A. M. 1989. Shelf-stable pineapple for long-term non refrigerated storage. *Lebensm.-Wess. M-Technol.* 22:233–236.

Argaiz, A., López-Malo, A., and Welti-Chanes, J. 1991. Conservación de frutas por factores combinados. 1. Papaya y Piña. *Boletín de Divulgación de los Grupos Mexicanos* 4, 9–17. Universidad de las Américas, Puebla, Mexico.

Ben-Arie, R., and Kislev, N. 1979. Ultrastructural changes in the cell walls of ripening apple and pear fruit. *Plant Physiol.* 64:197–202.

Bolin, H. R., Huxsoll, C. C., Jackson, R., and Ng, K. C. 1983. Effect of osmotic agents and concentration on fruit quality. *J. Food Sci.* 48:202–205.

Cerruti, P., Alzamora, S. M., and Chirife, J. 1990. A multiparameter approach to control the growth of *Saccharomyces cerevisiae* in laboratory media. *J. Food Sci.* 55:837–840.

Chirife, J., and Favetto G. 1992. Some physico-chemical basis of food preservation by combined methods. *Food Res. Intern.* 25:389–396.

Conway, J., Castaigne, F., Picard, G., and Vovan, X. 1983. Mass transfer consideration in the osmotic dehydration of apples. *Can. Inst. Food Sci. Technol. J.* 16:25–29.

Dixon, G. M., and Jen, J. J. 1977. A research note: Changes of sugar and acids of osmovac-dried apple slices. *J. Food Sci.* 42:1126.

Domingo, A. L., and Lluch, R. M. A. 1991. Estudio de las modificaciones microestructurales de la manzana *Malus comrnunis* L. 'Granny Smith' sometida a la deshidratación osmótica. In *Anales de Investigación del Master en Ciencias e Ingeniería de Alimentos.* Vol. 1, Fito, P.,

Serra, J., Hernández, E., and Vidal, D., eds. Consejo Nacional de la Universidad Politécnica de Valencia, Spain, pp. 709–732.

Farkas, D. F., and Lazar, M. E. 1969. Osmotic dehydration of apple pieces: Effect of temperature and syrup concentration on rates. *Food Technol.* 23:688–692.

Forni, E., Torreggiani, D., Battiston, P., and Polesello, A. 1986. Research into chances of pectic substances in apricot and peaches processed by osmotic dehydration. *Carbohydrate Polymers.* 6:379–393.

Glenn, G. M. and Poovaiah. B. W. 1990. Calcium-mediated post-harvest changes in texture and cell wall structure and composition in 'Golden Delicious' apples. *J. Amer. Soc. Hort. Sci.* 115:962–968.

Gould, G. W., and Jones, M. V. 1989. Combination and synergistic effects. In *Mechanisms of Action of Food Preservation Procedures,* Gould, G. W., ed. pp. 401–421. London: Elsevier Applied Science.

Johnson, D. S. 1979. New techniques in the post-harvest treatment of apple fruit with calcium salts. *Commun. Soil. Sci. Plant Anal.* 10:373–382.

Kanujoso, B. W. T., and Luh, B. S. 1967. Texture, pectin, and syrup viscosity of canned cling peaches. *Food Technol.* 21:457–460.

Karel, M. 1975. Dehydration of foods. In *Principle of Food Science. Part II. Physical Principles of Food Preservation.* Karel, M., Fennema, O. K., and Lund, D. E., eds. New York: Marcel Dekker, Inc.

Kovács, E., Keresztes, Á., and Kovács, J. 1988. The effect of gamma irradiation and calcium treatment on the ultrastructure of apples and pears. *Food Microstruct.* 7:1–14.

Lapsley, K. G., Escher, F. E., and Hoehn, E. 1992. The cellular structure of selected apple varieties. *Food Micro. Struct.* 11:339–349.

Le Maguer, M. 1988. Osmotic dehydration: Review and future directions. In *Proceedings of Symposium on Progress in Food Preservation Processes,* Vol. 1, pp. 283–309. CERIA, Brussels, Belgium.

Leistner, L. 1985. Hurdle technology applied to meat products of the shelf stable and intermediate moisture food types. In *Properties of Water in Foods.* Simatos, D. and Martinus Nijhoff J. L., eds. The Netherlands: Dordrecht. pp. 309–329.

Leistner, L. 1992. Food preservation by combined methods. *Food Res. Int.* 25:151–158.

Lerici, C. R., Pinnavaia, G., Dalla Rosa, M., and Bartolucci, L. 1985. Osmotic dehydration of fruit: Influence of osmotic agents during behavior and product quality. *J. Food Sci.* 50:1217–1219.

Levi, A., Gagel, S., and Juven, B. J. 1983. Intermediate moisture tropical fruit products for developing countries. I. Technological data on papaya. *J. Food Sci.* 18:667–685.

Levi, A., Gagel, S., and Juven, B. J. 1985. Intermediate-moisture tropical fruit products for developing countries. II. Quality characteristics of papaya. *J. Food Technol.* 20:163–175.

Marcotte, M. 1988. Mass transport phenomena in osmotic processes: Experimental measurements and theoretical considerations. M.S. thesis. The University of Alberta, Edmonton, Alberta, Canada.

Mason, J. L. 1976. Calcium concentration and firmness of stored 'McIntosh' apples increased by calcium chloride solutions plus thickener. *Hort. Sci.* 11:504–505.

Monsalve-González, A., Barbosa-Cánovas, G. V., and Cavalieri, R. P. 1993a. Mass transfer and textural changes during processing of apples by combined methods. *J. Food Sci.* 58(5):1118–1124.

Monsalve-González, A., Barbosa-Cánovas, G. V., Cavalieri, R. P., McEvily, A. J., and Iyengar, R. 1993b. Control of browning during storage of apple slices preserved by combined methods. 4-hexylresorcinol as anti-browning agent. *J. Food Sci.* 58:797–800.

Moy, J. H., Lav, N. B. H., and Dollar, A. M. 1978. Effect of sucrose and acids on osmovac-dehydration of tropical fruits. *J. Food Proc. Preserv.* 2:131.

Poovaiah, B. W. 1986. Role of calcium in prolonging storage life of fruits and vegetables. *Food Technol.* 40:86–89.

Raoult, A., Lafont, F., Rios, G., and Gilbert, S. 1989. Osmotic dehydration: A study of mass transfer in terms of engineering properties. In *Drying 89,* Mujumdar, A.S. and Roques, M., Eds. New York: Hemisphere Publishing Corporation, p. 487.

Rojas, R., Sauceda, A., and Avena, R. 1991. Puré de mango conservado mediante factores combinados. *Boletín de Divulgación de los Grupos Mexicanos.* 4:18–25. Universidad de las Américas, Puebla, Mexico.

Sams, C. E., and Conway, W. S. 1984. Effect of calcium infiltration on ethylene production, respiration rate, soluble polyuronide content, and quality of 'Golden Delicious' apple fruit. *J. Amer. Soc. Hort. Sci.* 109:53–57.

SAS Institute, Inc. 1985. *SAS® User's Guide: Statistics,* Version 5 Edition. Cary, NC: SAS Institute, Inc., Statistical Analysis System.

Stow, J. 1989. The involvement of calcium ions in maintenance of apple fruit tissue structure. *J. Exp. Bot.* 40:1053–1057.

Trakoontivakorn, G., Patterson M. E., and Swanson, B. 1988. Scanning electron microscopy of cellular structure of Granny Smith and Red Delicious apples. *Food Microstruct.* 7:205–211.

CHAPTER 15

Packaging and Shelf-Life Study of Apple Slices Preserved by Combined Methods Technology

EMILY H. REN
GUSTAVO V. BARBOSA-CÁNOVAS

INTRODUCTION

O NE of the greatest innovations in food technology is the optimization of water activity in foods in order to extend shelf life (Paulus, 1990). An increasing demand for minimally processed fruits has prompted researchers to study the prospects of "hurdle technology" or "combined methods technology" as a novel processing and storage technique that is simple and low cost (Monsalve-González et al., 1993a, 1993b, 1995; Willcox et al., 1994).

Combined methods technology is a nonthermal processing technique. It employs a combination of restriction factors, or hurdles, such as mild heating, water activity reduction, chilling, pH, redox potential change, preservatives, and competitive flora, to obtain microbial, physical, and chemical stability of food products. The products are "shelf-stable products," meaning they are storable without refrigeration and are healthier, tastier, and more economical to consumers (Leistner, 1992). The development of such products could also meet increasing demands for minimally processed, fresh-like food products and for out-of-season processing in many kinds of foods, such as confectionery, bakery foods, and dairy products (Alzamora et al., 1993).

Combined methods technology has been applied mainly to meat products (Leistner, 1987). More recently, some researchers have reported effective fruit preservation by combined methods technology, including Flora et al. (1979), Giangiacomo et al. (1987), Torreggiani et al. (1988), Aguilera et al. (1992), Alzamora et al. (1993), Farkas (1994), Knorr (1994), and López-Malo et al. (1994). Apple products treated by combined methods technology through osmotic dehydration and then freezing or vacuum drying have been reported by Pointing et al. (1972), Dixon et al. (1976), Hawkes and Flink (1978), Monsalve-González et al. (1993a, 1993b), and Quinteros-Ramos et al. (1993). Intermediate moisture food technology (IMF), which can also make fruit products stable at room temperature, is limited by sensory considerations, given the high solute concentrations required to lower water activity to a safe level (López-Malo et al., 1994).

Packaging method is also an important factor in extending the shelf life of food preserved by combined methods technology. The functions of food packaging were defined by the Codex Alimentarius Commission (1985) as "to preserve food quality and freshness, add appeal to consumers, and facilitate storage and distribution."

Alzamora et al. (1993) reported that combined methods technology was successfully applied to tropical and subtropical fruits, including peaches, mangos, papayas, pineapples, bananas, and chicozapotes. Their products were covered with enough syrup to remain stable in non-refrigerated storage for three to eight months. According to Buick and Damoglou (1987), vacuum packaging significantly extended the shelf life of "ready-to-use" sliced carrots when stored at 4°C for five to eight days.

In general, packaging methods for fruit and vegetable products demand flexibility in film form, a high barrier to moisture and oxygen, a barrier to flavor and odor transfer, easy sealability, and transparency for marketing appeal and identification (Paine and Paine, 1983; Brown, 1992; Swanson et al., 1994).

Nitrogen is the most frequently used inert gas for packaging applications. It is used in the gas flushing operation (Hirsch, 1991). It is inexpensive, has no color or odor, does not cause any changes in the packaged product, and is readily available.

Foil is the only choice available for high-barrier flexible packaging material. The transmission rate of both oxygen and moisture can be dramatically reduced with the aid of foil, but cannot reach zero. The need for foil should be determined by the shelf-life expectancy of the product and its maximum permissible moisture loss or gain, as well as oxygen susceptibility. In the food industry, foil packaging is used for soup mixes or other dry powdered products that are very sensitive to moisture and are stored at room temperature on an open shelf (Hirsch, 1991). Sometimes, just a portion of the package is covered with aluminum. This provides the package with product visibility while at the same time making advantageous use of the mirror-like appeal of a metalized composite.

The purpose of this research was to study the shelf-life of apple slices preserved by combined methods technology as a function of different types of packaging materials to meet the demand for storage at room temperature for several months.

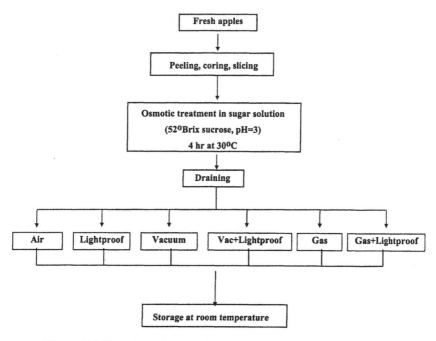

Figure 15.1 Flow-chart of preservation of apple slices by combined methods.

MATERIALS AND METHODS

PREPARATION OF SHELF-STABLE APPLE SLICES

Red Delicious apples were purchased from a local supermarket. The apples were peeled, cored, and sliced with an Apple Parer Corer & Slicer (Model No. 300, White Mountain Freezer Inc., Winchendon, MA). Apple slices 1-cm thick, were immediately immersed in an acidified sugar solution pH 3.0 at 30°C for 4 hours in a proportion of 3:1 (weight/weight). The process used for preparation of the apple slices is shown in Figure 15.1.

The acidified sugar solution was chosen to reduce the apples water activity to the equilibrium range of 0.93–0.97. The sugar solution also contained 0.7% malic acid (Bartek Chemical Company Ltd.) to adjust pH; 1,000 ppm sorbic acid (Eastman Chemical Company, Rochester, NY) to inhibit mold and yeast growth; 0.025% of 4-hexylresorcinol (Opta Food Ingredient Inc., Cambridge, MA) and 0.5% of ascorbic acid (EM Science, Germany) as anti-browning agents; and 0.5% calcium chloride (EM Science, Germany) to maintain firmness during the osmotic treatment.

PACKAGING AND STORAGE

After osmotic treatment, treated apple slices were drained and packed in transparent, low moisture and oxygen permeable 3-mil thick plastic vacuum pouches (Koch Supplies Inc., Kansas City, MO) and 3.6-mils thick light-proof aluminized polyester pouches (CAL VF4M, Caltex Plastics, Inc., Vernon, CA) with a moisture barrier. All bags were in retail packs (10 to 1,000 g). Three types of packaging environments—air, vacuum, and gas flush (nitrogen—and two types of packages—transparent and lightproof—were considered for each type of pouch. Then the treated apple slices were stored at room temperature (75 ± 1°F) for four months for the shelf-life study.

Fresh Red Delicious apple slices were used as control, and two kinds of commercial dried apple products were employed as reference materials: Mariani® premium dried apples (sulphur dioxide and/or sodium bisulfite were added as a preservative, and the apples need refrigeration after opening to maintain quality) and Townhouse® dried apple chunks (preserved with sulphur dioxide for color retention).

MICROBIOLOGICAL COUNTS

Microbiological analyses on apple slices were performed 24 hours after processing and then every month. Fresh and treated apple slices were homogenized in a blender using a 0.1% sterile peptone solution as the diluent. Microbial counts were also carried out in the fresh and used syrups. Aerobic mesophilics were counted using plate count agar that were incubated at 32°C for two days. Mold and yeast were isolated and counted using plate count agar with cloramphenicol and chlortetracycline incubated at 25°C for five days. Osmophilic yeast was recovered using the MY40 agar media and was incubated at 30°C for two days. Vacuum- and gas-packed samples were also incubated under anaerobic conditions. Microbiological counts were expressed as colony forming units per gram (cfu/g).

WATER ACTIVITY MEASUREMENT

The a_w of fresh and treated apple slices was measured using an Aqua Lab CX-2 electrical hygrometer (Decagon Devices, Inc., Pullman, WA). Samples were equilibrated at room temperature in the measurement containers for 30 min before measurements were taken. The a_w determinations were made in triplicate.

MOISTURE CONTENT

Moisture content (fresh basis) of the fresh and treated apple slices was determined by drying under vacuum (0.1 mm Hg) at 60°C to 70°C until constant weight was achieved (AOAC, 1984). Determinations were made in triplicate.

SOLUBLE SOLID AND TITRATABLE ACIDITY

Soluble solid content was determined from squeezed apple juice from slices. The soluble solid content was measured using a refractometer (Milton Roy Company, Rochester, NY). The pH was determined using an electronic pH meter (420A, Orion Research Inc.). For titratable acidity, apple slices were homogenized and titrated with 0.1N NaOH to an endpoint of pH 8.1. Titratable acidity was expressed as percent of malic acid.

TEXTURE MEASUREMENT

Texture of treated apple slices was measured according to Monsalve-González et al. (1993b). Apple cylinders were obtained from the fresh and treated apple slices using a cork borer. Twenty to 30 cylinders were compressed to 25% of their original height (6.5 to 5.5 mm) between two flat plates beyond rupture of the tissue. The compression/deformation force was applied to each cylinder specimen in a TA-XT2 Texture Analyzer (Texture Technologies Corp., Menlo Park, CA). In the compression mode, the TA-XT2 automatically measured the force required by the probe to travel the programmed distance at a preselected constant speed through the sample. The speed of the plate was set at 3 mm/min, and the average measurement of the peak force with the 95% confidence level was reported.

CALCIUM DETERMINATION

Calcium was determined by atomic absorption according to Conway et al. (1989) and Ferguson et al. (1979). For analysis, duplicate samples (200 to 250 mg) were weighed into small vials and were ashed in a furnace at 550°C for 24 hours. The ashes were mixed with 2,000 ppm Lanthanum and water, dissolved to result in 20 m to measure the calcium content by an atomic absorption spectrophotometer (Perkin Elmer, Model 5890).

RESIDUE ANALYSIS OF ASCORBIC ACID, SORBIC ACID, AND 4-HEXYLRESORCINOL

The HPLC system consisted of an E Lab Model 2020 gradient programmer and data acquisition system (OMS Tech., Miami, FL), an Eldex Model AA pump (Eldex Laboratories, Menlo Park, CA), and a Valco six-port injector (Valco Instrument, Houston, TX). The sample separation and detection employed a reversed-phase column (Bandapak, C18, 3.9 × 300 mm (Waters, Part 27324) and an HP 1040A diode array HPLC detector.

AA was determined by HPLC mainly according to Singh et al. (1983). AA was analyzed quantitatively by HPLC with the diode array detector at 242 nm. SA and 4HR were extracted and analyzed simultaneously using a modified method that was described by Torres et al. (1985). Accurately weighed 20-g treated apple slices, 5-g Celite, and 100-ml methanol were put in a small Waring blender jar. After mixing, the slurry was filtered through a Whatman No. 2 filter paper and the jar and the solid were rinsed with 50 ml of methanol, and combined the solutions were then diluted to 250 ml with methanol. The diluted solution was transferred to a separation funnel and was partitioned by the addition of 100 ml 0.5N NaOH and 50 ml of a 1:1 mixture of petroleum ether and ethyl ether mixture. The aqueous phase was transferred to a 1,000-ml Erlenmeyer. The organic layer was gently washed with 15 ml of 0.5 NaOH and then was discarded. The aqueous extracts were combined and titrated to pH 2.0 with HCl (1 + 1), and the total volume was accurately determined. A 200-ml aliquot was passed through a Millipore filter, and the filtered solution was quantitatively determined by HPLC with the diode array detector at 254 nm for SA and at 278 nm for 4HR. A two-solvent gradient elution was used to separate SA and 4HR. The weak solvent was 0.1% trifloracetic acid (TFA) in de-ionized water, and the strong solvent was 0.1% TFA in acetonitrile. Flow rate was 1 ml/min, and the sample was 50 μl.

COLOR MEASUREMENT

Color of the fresh and treated apple slices was measured using a Minolta CM-2002 Spectrophotometer (Minolta Camera Co., LTD, Chuo-Ku, Osaka 541, Japan). L^* and a^* values were measured on apple slices through plastic pouches at four places on the apple slices. Each pouch contained four slices from four different apples, and the averages of L^* and a^* were reported. Reported errors are calculated at a 95% confidence level.

SENSORY EVALUATION

Samples of apple slices and apple pies were subjected to an acceptability and multiple sample comparative test by an untrained sensory panel. Apple pies were prepared with either fresh apple slices or treated apple slices using the same formula and baking conditions. The pie filling was prepared by mixing 3/4 cup sugar, 3/4 cup all-purpose flour, 1/2 teaspoon ground nutmeg, 1/2 teaspoon ground cinnamon, a dash of salt, 6 cups thinly sliced pared tart apples (about 6 medium), and 2 tablespoons of butter (Betty Crocker's Cookbook). The premade pastry was bought from a local supermarket. Pies were covered with a top crust that had slits cut in it, were sealed and fluted, and were baked at 425°F for 40 to 50 min. Color (appearance), firmness, tartness, flavor, and preference (acceptability) of apple slices were evaluated on a hedonic scale of 9 points (9 = high or like extremely; 1 = low or dislike extremely). Sensory analyses were undertaken at time zero and at one and four months of storage. Samples were presented to each panelist at each session in a coded paper plate under white light to let each judge evaluate the appearance of samples. Fresh apple slices and apple pies baked with fresh apple slices were used as control and reference standards. Sensory evaluation data represents the results of 40 to 60 untrained judges.

STATISTICAL ANALYSIS OF RESULTS

This experiment was designed as a three-way factorial, with package, package environment, and storage time as the main factors. Two batches of treated apples were used as replicates. Data were evaluated by analysis of variance and least square means calculated by the General Linear Models Procedure (PROC GLM) of the Statistical Analysis System (SAS Institute, Inc., 1987). Data were evaluated for significant difference at a 95% confidence level.

RESULTS AND DISCUSSION

OSMOTIC TREATMENT OF APPLE SLICES

The soluble solid content of apple slices increased from 12° Brix on fresh apples to 29 to 30° Brix after the osmotic treatment. This increase in soluble solid content was accompanied with a decrease in moisture content from 86.2% (kg water/100 kg dry solids) to 57.9–62.3% in the treated slices. The pH decreased from 4.04 to 3.69, and the titratable acidity (as % malic acid) increased from 0.40% to 0.82%.

PACKAGED APPLE SLICES

Apple slices treated by combined methods technology, packaged in transparent pouches with air, vacuum, and gas (nitrogen) flush, and stored at room temperature showed microbial counts of less than 10 cfu/g during 136 days of storage for yeast and mold, osmophilic yeast, and aerobic/anaerobic mesophilics. A pH of 3.69 on apple slices could prevent the growth of microorganisms.

Titratable acidity of apple slices packaged in transparent pouches increased in a steady pattern for each of the packaging techniques, as shown in Table 15.1. There was no significant difference between packaging techniques in the increase of titratable acidity. A decrease in pH was observed for each of

TABLE 15.1. The Titratable Acidity (% Malic Acid) of Fresh and Treated Apple Slices in Different Packaging in Transparent Plastic Pouches during Storage.

Time	Titratable Acidity (% Malic Acid)[*]				pH[*]			
(Months)	Fresh	Air	Vac	Gas	Fresh	Air	Vac	Gas
0	0.40[b]	0.82[a]	0.82[a]	0.82[a]	4.04[a]	3.69[b]	3.69[b]	3.69[b]
1		1.23[a]	1.15[a]	1.55[a]		3.46[a]	3.92[a]	3.04[a]
2		1.60[a]	1.58[a]	1.69[a]		2.67[a]	2.94[a]	2.61[a]
4		1.77[a]	1.64[a]			2.64[a]	2.88[a]	

[*]Air = air package, Vac = vacuum package, Gas = gas package, Fresh = fresh apple.
[a,b] Means within the same row having the same superscript are not significantly different ($P > 0.05$).

the packaging techniques with no significant difference between each technique. The decrease in pH may be explained in terms of sucrose hydrolysis, as discussed by López-Malo et al. (1994).

ASCORBIC ACID, SORBIC ACID, AND 4-HEXYLRESORCINOL RETENTION

The content of 4-hexylresorcinol (4HR), sorbic acid (SA), and ascorbic acid (AA) on apple slices after 3.5 months of storage varied significantly as a function of packaging technique (Table 15.2). 4HR content decreased most significantly in those pouches containing air (transparent and lightproof) with

TABLE 15.2. 4-Hexylresorcinol (4HR), Sorbic Acid (SA), and Ascorbic Acid (AA) Content Changes of Apple Slices Preserved by Combined Methods in Different Packages.

	Time (Month)	Fresh Apple	Treated Apple[*]					
			Air	Vac	Gas	V + L	G + L	L
4HR	0	0	105	105	105	105	105	105
(ppm)	3.5	0	32[c]	43[b]	49[a]	50[a]	51[a]	36[c]
Loss (%)			69.5	59.0	53.3	52.4	51.4	65.7
SA	0	0	705	705	705	705	705	705
(ppm)	3.5	0	384[d]	610[b]	607[b]	644[a]	633[a]	528[c]
Loss (%)			45.5	13.5	13.9	8.7	10.2	25.1
AA	0	5	120	120	120	120	120	120
(mg/100g)	3.5	ND	0[c]	1[b]	1[b]	4[a]	3[ab]	0[c]
Loss (%)			100	99.2	99.2	96.7	97.5	100

[*] Air = air package, Vac = vacuum package, Gas = gas package, L = lightproof package, V + L = vacuum + lightproof package, G + L = gas + lightproof package.
[a,b,c,d] Means within the same row having the same superscript are not significantly different ($P > 0.05$).

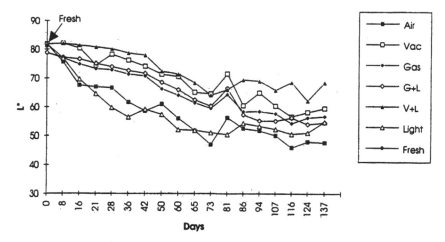

Figure 15.2 L* value of apple slices preserved by combined methods in different packages during storage. Air = air package, Vac = vacuum package, Gas = gas package, Light = lightproof package, V+L = vacuum + lightproof package, G + L = gas + lightproof package. Average measurement of the peakforce with 95% confidence level was reported.

a total loss of 65.7 to 69.5%. Meanwhile, there was no significant difference between lightproof pouches packaged under vacuum or nitrogen flush and transparent pouches packaged with nitrogen where total loss of 4HR ranged from 51.4 to 53.3%. The decrease in ascorbic acid was significant for all six packaging techniques considered in this study. The concentration of AA after 3.5 months ranged from 0 to 4 ppm/100 g of product when initially there was 120 ppm/100 g of product. The reduction of 4HR and AA may be explained in terms of the role of those compounds in preventing enzymatic browning (Mayer and Harel, 1979; Lambrecht, 1995). The change in color parameters $L*$ and $a*$ is shown in Figures 15.2 and 15.3. Those treatments containing air showed a rapid change in both L^4 and a^4 at the beginning of the storage period and also presented higher reduction in 4HR after 3.5 months. The other four packaging alternatives (no air in the package) showed a lower rate of color change at the beginning of the storage period with a lower loss of 4HR after 3.5 months. Oxygen is required for the formation of *o*-quinones during enzymatic browning. 4HR participates in the chemical reaction early in the formation of diphenoles, which explains the partial comsumption of 4HR. Meanwhile, AA acts as a reducing agent that reacts with *o*-quinones (colored compounds) to produce diphenols (colorless compounds). The continuous formation of *o*-quinones leads to total consumption of AA. Those packages containing oxygen favored enzymatic browning while the other four packaging techniques reduced the suceptibility of apple slices to browning.

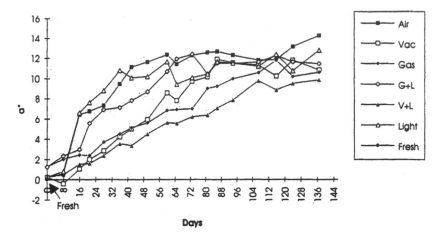

Figure 15.3 a* Value of apple slices preserved by combined methods in different packages during storage. Air = air package, Vac = vacuum package, Gas = gas package, Light = lightproof package, V + L = vacuum + lightproof package, G + L = gas + lightproof package. Average measurement of the peakforce with 95% confidence level was reported.

The loss of sorbic acid (SA) seemed to be enhanced with the presence of oxygen and its exposure to light. The higher loss in SA occurred in those packages containing oxygen followed by those exposed to light (Table 15.2).

Texture analyses showed a significant decrease in peak force of apple slices as a function of storage time when packaged in the presence of air and exposed to light. Lack of air (using vacuum or nitrogen flush) preserved the texture of treated apple slices as shown in Figure 15.4.

SENSORY EVALUATION OF APPLE SLICES AND APPLE PIES

The texture of apple slices was negatively affected by the osmotic treatment, as shown in Figure 15.4 and as noted by the sensory panel (Table 15.3). Meanwhile, there was an increase in sweetness because of the increase in the content of soluble solids. The increased value of tartness observed by the sensory panel may be attributed to the 4HR, AA, and SA content of the apple slices. The sensory panel found no differences between the apple flavor of fresh non-treated apple slices and those exposed to the osmotic treatment; similar results were obtained for the overall acceptance of the treated product.

Once the apple slices were used in the preparation of apple pies, the texture of treated apple slices obtained a better score than fresh apple slices (Table 15.3). The improved texture may be explained in terms of synergistic interaction of 4HR, AA, and SA with the apple while baking the pies, which seems to protect the texture of the slices.

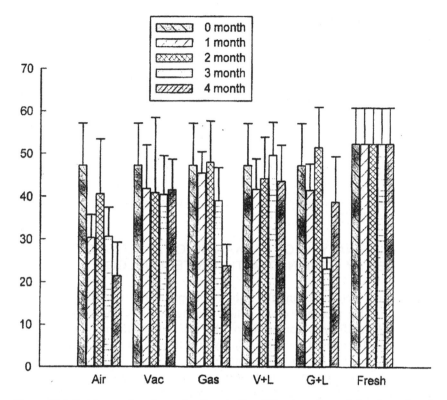

Figure 15.4 Peakforce values for apple slices in five different packages during 4 months of storage. Air = air package, Vac = vacuum package, Gas = gas package, V + L = vacuum + lightproof package, G + L = gas + lightproof package. Average measurement of the peakforce with 95% confidence level was reported.

The sensory panel found a decreased acceptability of apple slices packaged with air along the four months of storage. Meanwhile, apple slices packaged using vacuum or nitrogen flush maintained an acceptance level similar to that of fresh apple slices (Tables 15.4 and 15.5). A similar pattern was observed for flavor and color along the four months of storage.

The sensory panel found no significant differences among packaging methods for most of the sensory characteristics on treated apple slices stored at room temperature for one to two weeks and used in the preparation apple pies. Similar results were obtained for apple pies prepared with slices stored for five to six weeks. Nevertheless, a steady decrease in the scores for each sensory characteristic was observed (Table 15.6), indicating that the sensory panel detected the deterioration of the apple slices. Firmness, flavor, and overall acceptance of apple slices packaged with air obtained lower scores from the sensory panel.

TABLE 15.3. Sensory Evaluation* of Fresh and Preserved Apple Slices and Apple Pies Made from Fresh and Preserved Apple Slices after 0 Months of Storage.

		Texture	Sweet	Tart	Flavor	Acceptance
Apple slices	Fresh	6.58[a]	5.43[b]	3.91[b]	5.77[a]	5.83[a]
	Treated	4.72[b]	6.60[a]	4.59[a]	5.21[a]	5.28[a]
	Std Err	0.23	0.21	0.19	0.24	0.23
Apple pies	Fresh	4.32[b]	5.67[a]	3.68[b]	5.18[a]	5.91[a]
	Treated	5.89[a]	6.46[a]	4.94[a]	5.28[a]	5.62[a]
	Std Err	0.19	0.19	0.17	0.19	0.19

* Samples were evaluated using a 9-point scale, with 9 = like extremely and 1 = dislike extremely.
[a,b] Means within the same column having the same superscript are not significantly different ($P > 0.05$). This experiment was conducted one week after processing, while packages did not have influence on the quality of treated apple slices.

OVERALL ANALYSIS OF PACKAGING OF APPLE SLICES TREATED BY COMBINED METHODS TECHNOLOGY

The main factor affecting the stability of apple slices treated by combined methods is the presence of air when packaged. The quality of the slices decreased significantly compared to other packaging alternatives (gas flush, vacuum, and lightproof package). Meanwhile, there are no advantages in using lightproof pouches for packaging apple slices treated by combined methods technology. The most important hurdle during the packaging step is the removal of air from the pouch to improve the shelf stability of the product.

CONCLUSIONS

Apple slices processed by combined methods and packaged either under

TABLE 15.4. Sensory Evaluation* of Fresh and Preserved Apple Slices after 1 Month of Storage.

Package**	Color	Texture	Sweet	Tart	Flavor	Acceptance
Fresh	4.33[b]	5.96[a]	4.33[b]	3.51[b]	5.23[a]	5.29[a]
Vac	6.81[a]	5.01[b]	5.36[a]	4.44[a]	5.03[ab]	4.88[ab]
G + L	6.77[a]	4.90[b]	5.20[a]	4.15[a]	4.73[ab]	4.95[ab]
Gas	6.80[a]	4.63[b]	4.99[a]	4.43[a]	4.85[ab]	4.73[ab]
V + L	6.99[a]	4.89[b]	5.39[a]	4.28[a]	4.78[b]	4.59[ab]
Air	4.59[b]	3.83[c]	4.62[b]	4.26[a]	4.82[ab]	4.54[b]
Std Err	0.17	0.16	0.17	0.13	0.16	0.16

*Samples were evaluated using a 9-point scale, with 9 = like extremely and 1 = dislike extremely.
**Air = air package, Vac = vacuum package, Gas = gas package, L = lightproof package, V + L = vacuum + lightproof package, G + L = gas + lightproof package.
[a,b,c] Means within the same row having the same superscript are not significantly different ($P > 0.05$).

TABLE 15.5. Sensory Evaluation* of Apple Slices in Different Packaging Methods** after 4 Months of Storage and Comparison of Different Samples and Packages.

Package	Color	Texture	Flavor	Acceptance	Advantages	Disadvantages
Air	2.08^d	3.43^b	3.93^b	3.11^c	Can see-through, simple	Short shelf life (1 to 2 months)
Vac	5.40^{ab}	5.15^a	5.18^a	4.83^a	Protect color and flavor	Sometimes lose appearance
Gas	5.12^{ab}	4.80^a	4.79^a	4.54^a	Protect color and flavor	Need gas flushing operation
V + L	5.62^a	5.10^a	5.12^a	4.92^a	Provide optimum protection of flavor and against browning	Cannot see through
G + L	5.89^a	5.01^a	4.87^a	4.88^a	Protect color and flavor	Need gas flushing operation
Mariani	5.41^a	4.78^a	4.02^b	3.89^b	Good appearance	Strong sulfite flavor
Townhouse	3.86^c	4.34^{ab}	4.11^b	3.97^b	Good flavor	Appearance is not good
Std Err	0.14	0.16	0.15	0.16		

*Samples were evaluated using a 9-point scale, with 9 = like extremely and 1 = dislike extremely.
**Air = air package, Vac = vacuum package, Gas = gas package, V + L = vacuum + lightproof package, G + L = gas + lightproof package.
a,b,c,d Means within the same column having the same superscript are not significantly different ($P > 0.05$).

TABLE 15.6. Sensory Evaluations* of Treated Apple Slices in Apple Pies in Different Packages** during the First 2 Months of Storage at Room Temperature.

	Weeks	
Attribute	1–2	5–6
Firmness		
Air	5.89[ax]	5.05[by]
Vac	5.89[ax]	5.60[ax]
Gas	5.89[ax]	5.41[by]
V + L	5.89[ax]	5.70[ax]
G + L	5.89[ax]	5.36[by]
Sweetness		
Air	6.46[ax]	5.56[ay]
Vac	6.46[ax]	5.71[ay]
Gas	6.46[ax]	5.74[ay]
V + L	6.46[ax]	5.56[ay]
G + L	6.46[ax]	5.58[ay]
Tartness		
Air	4.94[ax]	4.56[ax]
Vac	4.94[ax]	4.48[ax]
Gas	4.94[ax]	4.38[ay]
V + L	4.94[ax]	4.17[ay]
G + L	4.94[ax]	4.42[ay]
Flavor		
Air	5.28[ax]	4.94[ax]
Vac	5.28[ax]	5.17[ax]
Gas	5.28[ax]	5.22[ax]
V + L	5.28[ax]	5.02[ax]
G + L	5.28[ax]	5.19[ax]
Acceptance		
Air	5.62[ax]	4.80[by]
Vac	5.62[ax]	4.97[ay]
Gas	5.62[ax]	5.12[ax]
V + L	5.62[ax]	4.88[aby]
G + L	5.62[ax]	5.30[ax]

*Samples were evaluated using a 9-point scale, with 9 = like extremely and 1 = dislike extremely.
**Air = air package, Vac = vacuum package, Gas = gas package, V + L = vacuum + lightproof package, G + L = gas + lightproof package.
[a,b] Means within the same column having the same superscript are not significantly different ($P > 0.05$).
[x,y] Means within the same row having the same superscript are not significantly different ($P > 0.05$).

vacuum or nitrogen flush retain their fresh-like quality, including light yellow color, natural flavor, and firm texture during storage at room temperature. This product is microbiologically and organoleptically stable for at least four months of storage. The studied packaging alternatives for apples treated by combined methods allow for immediate consumption with high acceptability. They can be managed as both final products and as food ingredients.

Combined methods technology is a simple and inexpensive alternative to refrigeration, freezing, and other processing methods intensive in energy expenditure and capital investment. Sensory evaluation of the treated apple slices substituted in apple pies corresponded to high acceptability and preference. The development of such a fresh-like, wholesome, and high-quality fruit product can provide additional uses for the fruit. Therefore, food produced by combined methods technology is geared to the increasing demand for products that allow greater creativity in the household kitchen, such as incorporation into pies, salads, and pizza. This technology is an important alternative to traditional canning and drying, which stabilize products but dramatically change the organoleptic properties of apple.

ACKNOWLEDGEMENT

The authors wish to thank Mr. Humberto Vega-Mercado for his aid with editing work.

REFERENCES

Aguilera, J. M., Francke, A., Figueroa, G., Bornhardt, C., and Cifuentes, A. 1992. Preservation of minced pelagic fish by combined methods technology. *Inter. J. Food Sci. and Technol.* 27:171–177.

Alzamora, S. M., Tapia, M. S., Argaiz, A., and Welti-Chanes, J. 1993. Application of combined methods technology in minimally processed fruits. *Food Res. Int.* 26:125–130.

AOAC. 1984. *Official Methods of Analysis of the Association of Official Analytical Chemists,* 14th Edition. Arlington, VA: Association of Analytical Chemists, Inc.

Betty Crocker's Cookbook. New and revised edition. Minneapolis, MN: Golden Inc. p. 295.

Brown, W. E. 1992. *Plastics in Food Packaging: Properties, Design, and Fabrication.* New York: Marcel Dekker, Inc.

Buick, P. K., and Damoglou, A. P. 1987. The effect of vaccum packaging on the microbial spoilage and shelf life of "ready-to-use" sliced carrots. *J. Sci. Food Agric.* 38:167–175.

Codex Alimentarius. 1985. Food and Agriculture Organization of the United Nations, Rome, Italy.

Conway, W. S., Gross, K. C., Boyer, C. D., and Sams, C. E. 1989. Inhibition of penicillin expansion polygalacturonase activity by increased apple cell wall calcium. *Phytopath.* 78:1052.

Dixon, G. M., Jen, J. J., and Paynter, V. A. 1976. Tasty apple slices result from combined osmotic-dehydration and vacuum drying process. *Food Prod. Dev.* 10 (7):60–64.

Farkas, J. 1994. Combination of mild preservative factors to improve safety and keeping quality of chilled foods. In *Minimal Processing of Foods and Process Optimization: An Interface* R. Paul Singh and Fernanda A. R. Oliveira, eds. Boca Raton, FL: CRC Press, pp. 43–56.

Ferguson, J. B., Reid, M. S., and Prasad, M. 1979. Calcium analysis and the prediction of bitter pot in apple fruit. *N.Z.J. Agric. Res.* 22:485.

Flora, L. F., Beuchat, L. R., and Rao, V. N. M. 1979. Preparation of a shelf-stable, intermediate moisture food product from muscadine grape skins. *J. Food Sci.* 44:854–856.

Giangiacomo, R., Torreggiani, D., and Abbo, E. 1987. Osmotic dehydration of fruit. 1. Sugars exchange between fruit and extracting syrups. *J. Food Proc. Preserv.* 11 (3):183–195.

Hawkes, J., and Flink, J. M. 1978. Osmotic concentration of fruit slices prior to freeze dehydration. *J. Food Proc. Presev.* 2:265–284.

Hirsch, A. 1991. *Flexible Food Packaging: Questions and Answers.* New York: Van Nostrand Reinhold.

Knorr, D. 1994. Nonthermal processing for food preservation. In *Minimal Processing of Foods and Process Optimization: An Interface.* R. Paul Singh and Fernanda A. R. Oliveira, eds. Boca Raton, FL: CRC Press, pp. 3–16.

Lambrecht, H. S. 1995. Sulfite substitutes for the prevention of enzymatic browning in foods. In *Enzymatic Browning and Its Prevention,* C. Y. Lee and J. R. Whitaker, eds. ACS Symposium Series 600. American Chemical Society. pp. 313–323.

Leistner, L. 1987. Shelf stable product and intermediate moisture foods based on meat. In *Water Activity: Theory and Applications to Food,* L. B. Rockland and L. Beuchat, eds. New York: Marcel Dekker, Inc. pp. 295–327.

Leistner, L. 1992. Food preservation by combined methods technology. *Food Res. Inter.* 25 (2):151–158.

López-Malo, A., Palou, E., Welti, J., Corte, P., and Argaiz, A. 1994. Shelf-stable high moisture papaya minimally processed by combined methods technology. *Food Res. Int.* 27:545–553.

Mayer, M. A., and Harel, E. 1979. Review: Polyphenol oxidase in plants. *Photochem.* 18:193.

Monsalve-González, A., Barbosa-Cánovas, G. V., and Cavalieri, R. P. 1993a. Control of browning during storage of apple slices preserved by combined methods technology. 4-hexylresorcinol as anti-browning agent. *J. Food Sci.* 58 (4):797–800, 826.

Monsalve-González, A., Barbosa-Cánovas, G. V., and Cavalieri, R. P. 1993b. Mass transfer and textural changes during processing of apples by combined methods technology. *J. Food Sci.* 58 (5):1118–1124.

Monsalve-González, A., Barbosa-Cánovas, G. V., McEvily, A. J., and Iyengar, R. 1995. Inhibition of enzymatic browning in apple products by 4-hexylresorcinol. *Food Technol.* April:110–118.

Paine, F. A., and Paine, H. Y. 1983. *A Handbook of Food Packaging.* Glasgow: Leonard Hill.

Paulus, K. 1990. Necessary actions of the industry for successful food technology in the year 2000. In *Food Technology in the Year 2000.* S. Lindroth and S. S. I. Ryynanen, eds. No. 47, pp. 67–86. Bibl. Nutr. Dieta, Basel, Karger.

Pointing, J. D., Jackson, R., and Watters, G. 1972. Refrigerated apples: Preservative effects of ascorbic acid, calcium and sulfites. *J. Food Sci.* 37:434–436.

Quintero-Ramos, A., De la Vega, C., Hernández, E., and Anzaldúa-Morales, A. 1993. Effect of the conditions of osmotic treatment on the quality of dried apple discs. *AIChE Symp. Series 297. Food Dehydration* 89:108–113.

SAS Institute, Inc. 1987. Release 6.04. Cary, NC: SAS Institute.

Singh, R. K., Lund, D. B., and Buelow, F. H. 1983. Storage stability of intermediate moisture apples: Kinetic of quality. *J. Food Sci.* 48:939.

Swanson, B. G., Berrios, J. D. J., and Patterson, M. E. 1994. Modified atmosphere packaging of fresh fruits: New control techniques. *IFT* annual meeting, Chicago, IL.

Torreggiani, D., Forni, F. E., and Rizzolo, A. 1988. Osmotic dehydration of fruit. 2. Influence of the osmosis time on the stability of processed cherries. *J. Food Proc. Preserv.* 12 (1):27–44.

Torres, J. A., Motoki, M., and Karel, M. 1985. Microbial stabilization of intermediate moisture food surfaces. I. Control of surface preservative concentration. *J. Food Proc. Preserv.* 9:75.

Willcox, F., Hendrickx, M., and Tobback, P. 1994. The influence of temperature and gas composition on the evaluation of microbial and visual quality of minimally processed endive. In *Minimal Processing of Foods and Process Optimization: An Interface*. R. Paul Singh and Fernanda A. R. Oliveira, eds. Boca Raton, FL: CRC Press, pp. 475–492.

CHAPTER 16

Browning of Apple Slices Treated with Polysaccharide Films

NICOLÁS BRÁNCOLI
TERRY BOYLSTON
GUSTAVO V. BARBOSA-CÁNOVAS

INTRODUCTION

MINIMAL processing is defined as the handling, preparation, packaging, and distribution of agricultural commodities in a fresh-like state (Shewfelt, 1987). Problems such as enzymatic browing and microbial growth are a concern when using this process, especially for apples. Enzymatic browning in apples is caused by the enzyme polyphenoloxidase (PPO), also called orthodiphenol oxidase or catecholase (Nicolas et al., 1994). Polyphenoloxidase is one of the most important color deteriorative enzymes in fruits and vegetables (Uhlemann, 1993). In apples, PPO catalyzes the oxidation of phenolic compounds containing two o-dihydroxy groups to the corresponding o-quinone (Nicolas et al., 1994). Condensation of the o-quinones results in the formation of dark brown polymers called melanin. To inactivate PPO, refrigeration alone is insufficient. Although refrigeration temperatures slow enzymatic browning, they do not stop it from occurring. One promising way to inhibit PPO is the application of edible polysaccharide films because they are effective gas barriers and have little impact on the fresh taste of apple slices (Guilbert, 1986). Application of an edible film formulated with FDA-approved ingredients with desirable physical, sensory, and microbial properties to minimally processed apples could reduce the development of enzymatic browning. Among the possible approved ingredients, ascorbic acid proves to be a good reducing agent for oxidized polyphenolic substrates because it prevents polymerization, which causes the discoloration (Uhlemann, 1993). Potassium sorbate functions as an antimicrobial preservative to retard the growth of yeasts, molds, and bacteria. To maintain or improve the texture of apple slices, calcium chloride can be incorporated into the film. The ideal edible film should create a barrier that can retard the loss of desirable flavor volatiles and moisture while restricting the exchange of oxygen and carbon dioxide. Respiration and ethylene production should also be inhibited. The film should not create an anaerobic environment, as this would lead to anaerobic respiration, growth of anaerobic microbes, and an undesirable product. Research has shown that edible films protect apple pieces from moisture loss and oxidative browning for up to three days (Baldwin et al., 1995). The specific objectives of this research were to evaluate color changes in apple slices coated with a polysaccharide edible film by identifying instrumental measurements and sensory evaluation correlations on apple slices' color changes.

MATERIALS AND METHODS

FRUIT

Washington Red Delicious apples were washed, sliced, and cored. Slicing was accomplished with a stainless-steel slicer to obtain eight uniform apple slices. A total of 80 apples were used, 40 in each of two replications or lots. Lots were subdivided into four groups each (coated and uncoated held at 20°C, and coated and uncoated held at 4°C) to study changes in color at different storage temperatures. In each group, 10 apples were used for measuring instrumental and sensory color changes.

POLYSACCHARIDE FILM

Apple slices (25 g) were dipped in a polysaccharide film solution consisting of 10 g maltodextrin, 5 g methylcellulose, 8 ml glycerol, 2.5 g ascorbic acid, 1 g potassium sorbate, and 1 g calcium chloride dissolved in 1,000 g of distilled water. Coated slices were allowed to dry for 2 min and then were packaged into plastic bags (nylon/polyethylene, 3MIL) and sealed. Untreated apple slices were packaged into plastic bags and sealed immediately after slicing to serve as controls. Treated and untreated apple slices were stored at room temperature (20°C) and refrigeration temperature (4°C).

COLOR MEASUREMENT

Color was measured instrumentally using a Minolta® colorimeter (Model 2002, Minolta Camera Co. Ltd., Japan) Based on a Hunter modified L^*, a^*, b^* (CIE) system, $L^* = 100$ was white, $L^* = 0$ black, $a^* > 0$ red, $a^* < 0$ green, $b^* > 0$ yellow, and $b^* < 0$ blue. L^*, a^*, and b^* measurements were converted into the whiteness index (WI) according to $100 - [(100 - L^*)^2 + a^{*2} + b^{*2}]^{1/2}$

(Judd and Wyszecki, 1963). Lightness ($L*$) and the WI were then used for correlations to the sensory data to determine the appropriateness of their use as indicators of whiteness in treated and untreated apple slices. The length of the study was 12 days, with evaluations of the apple slices after 3, 5, 10, and 12 days in storage. A total of four measurements were made for each treatment.

SENSORY EVALUATION

Sensory evaluation was conducted using a 15-cm unstructured line scale with anchor points 1.5 cm from the ends labeled "brown" and "white." The scale assigns a value of 15 for white apple slices as seen in a fresh cut apple, and 1 for brown apple slices allowed to undergo extensive browning reactions. Panel training consisted of a discussion session about the definitions to be used for the terms white and brown in this study, as well as a display of fresh cut and brown apple slices to be used as references. Reference food samples were also present during all sensory evaluations. Eleven to 14 panelists evaluated apple slices at each session.

STATISTICAL ANALYSIS

Differences in both instrumental and sensory color characteristics and their correlation were statistically analyzed using ANOVA (analysis of variance). All analyses employed SAS computer software.

RESULTS AND DISCUSSION

COLOR CHANGES

The browning of apple slices was significantly ($P < 0.05$) inhibited during the 12 days of this study by storage at refrigeration temperature and to an even greater extent by coating apple slices in a polysaccharide film. As shown in Table 16.1, lightness ($L*$) values ranged from a high of 80.30 ± 1.01 for coated apple slices stored at 4°C for three days to a low of 64.61 ± 1.01 for uncoated apple slices stored at 20°C for 12 days. The WI followed this same trend with a range of 67.16 to 48.73. This trend can be explained by the suppression of the oxygen necessary for the enzymatic browning reaction and the low temperature. Indeed, PPO not only requires oxygen (Nicolas et al., 1994), but its related activity increases with temperatures between 4°C and 37°C (Uhlemann, 1993).

The sensory evaluation of apple slices demonstrated a clear distinction between uncoated and coated apple slices, but was not as effective as instrumental measurement for detecting the more subtle color differences observed

TABLE 16.1. Color Parameters of Uncoated and Polysaccharide Coated Apple Slices at 4°C and 20°C.

Instrumental Value/Treatment		Storage Time Days			
		3	5	10	12
L^* values					
Coated	4°C	$80.3 \pm 0.8^{a(a)}$	$79.5 \pm 0.7^{a(a)}$	$77.5 \pm 0.9^{b(a)}$	$76.6 \pm 1.0^{b(a)}$
	20°C	$78.0 \pm 1.9^{a(b)}$	$77.4 \pm 0.8^{a(b)}$	$76.0 \pm 1.0^{b(a)}$	$75.7 \pm 0.9^{b(a)}$
Uncoated	4°C	$71.4 \pm 1.0^{a(c)}$	$71.0 \pm 0.9^{a(c)}$	$67.6 \pm 1.3^{b(b)}$	$66.1 \pm 1.6^{c(b)}$
	20°C	$70.5 \pm 1.1^{a(c)}$	$69.9 \pm 1.6^{a(c)}$	$67.8 \pm 1.8^{b(b)}$	$64.6 \pm 1.0^{c(c)}$
WI					
Coated	4°C	$67.1 \pm 2.0^{a(a)}$	$67.0 \pm 2.7^{a(a)}$	$65.6 \pm 1.5^{b(a)}$	$63.8 \pm 3.2^{b(a)}$
	20°C	$66.0 \pm 3.8^{a(a)}$	$63.1 \pm 2.2^{b(b)}$	$63.8 \pm 1.3^{b(b)}$	$61.9 \pm 1.2^{c(a)}$
Uncoated	4°C	$54.3 \pm 0.9^{a(b)}$	$56.5 \pm 3.1^{b(c)}$	$53.3 \pm 1.2^{a(c)}$	$52.6 \pm 2.5^{c(b)}$
	20°C	$52.3 \pm 2.3^{a(c)}$	$51.9 \pm 2.1^{a(d)}$	$50.9 \pm 1.8^{b(c)}$	$48.7 \pm 1.8^{b(c)}$

[a,b,c] Means of two replications of four samples ±SD. Different letters in the same row indicate significant differences, $P < 0.05$. Different letters (parentheses) in the same column indicate significant differences, $P < 0.05$.

after storage temperature at days three and five. A tristimulus colorimeter to assay objective color differences was proposed by Kostyla and Clysdesdale (1978) to establish a correlation with consumer preferences. As shown in Table 16.2, samples scores ranged from 11.7 ± 0.83 for coated apple slices stored at 4°C for 10 days to a high of 3.2 ± 1.57 for uncoated apple slices stored at 20°C for 12 days.

In general, the experimental use of edible polysaccharide coatings for lightly processed fresh produce is in its early stages. Therefore, there is little information available on the use of edible films for this particular application. Among the different edible film formulations used in the past to inhibit the discoloration of apple slices, Uhlemann (1993) used different alginate-carragenan combinations resulting in a delay of the browning process for up to three days in coated apple slices compared to only 1 hour in untreated apple slices. Also, a soybean edible film coating developed at the U.S. Dept. of Agriculture (USDA), Agricultural Research Service preserved apple slice freshness in preliminary studies (Kinzel, 1992). Using a composite coating of chitosan and lauric acid, Pennisi (1992) inhibited both browning and water loss in cut apples. A composite coating of alginic acid, casein, and lipids that crosslink with the addition of calcium ions was also reported to reduce water loss and browning in cut apples (Wong et al., 1994).

CORRELATION ANALYSIS

There was a significant positive correlation at $P < 0.05$ between the sensory and instrumental evaluations of browning discoloration in the apple slices. An equally strong correlation was found between the lightness (L^*) values and sensory data ($r = 0.87$) and WI values and sensory data ($r = 0.86$) as shown in Figure 16.1. The use of correlation coefficients (r) between sensory attributes and objective techniques has been widely investigated. For example, Kader et al. (1977) observed significant correlations between titratable acidity and sensory panel assessments of sourness. Bartolome et al. (1996) evaluated pineapple slices in frozen storage and obtained correlation coefficients (r) ranked from 0.51 to 0.93 depending on the freezing method utilized. Avena-Bustillos et al. (1993) obtained a correlation coefficient of $r = 0.64$ in comparisons of WI values to sensory data. They also concluded that a properly performed sensory analysis could be even more sensitive and reliable for assessing white blush conditions. Resurreccion and Shewfelt (1985) reported that among 19 different objective and sensory measurements, color and firmness parameters gave the best correlation. Because quality evaluation of fruit by trained sensory panels based on statistical processes may be slow and requires many people, this method is not suited for daily quality control. Instrumental tests are needed that will give equally good results as sensory evaluations, but rapidly and with fewer people (Bartolome et al., 1996). This

TABLE 16.2. Sensory Evaluation [1 to 15, Brown (1) to White (15)] Scale of Uncoated and Polysaccharide Coated Apple Slices at 4°C and 20°C.

Sensory Value (1 to 15)/Treatment		Storage Time, Days			
		3	5	10	12
Color					
Coated	4°C	11.1 ± 1.6[a(a)]	10.0 ± 1.8[b(a)]	11.7 ± 1.8[a(a)]	10.6 ± 1.6[b(a)]
	20°C	10.0 ± 2.1[a(b)]	11.0 ± 1.3[b(b)]	11.1 ± 1.6[a(b)]	9.9 ± 1.8[b(b)]
Uncoated	4°C	5.7 ± 1.9[a(c)]	5.1 ± 2.4[b(c)]	5.6 ± 1.6[b(c)]	6.6 ± 1.7[c(c)]
	20°C	5.9 ± 1.7[a(c)]	7.2 ± 1.5[b(d)]	5.1 ± 1.5[a(c)]	3.2 ± 1.5[c(d)]

a,b,c Means of two replications of four samples ±SD. Different letters in the same row indicate significant differences, $P < 0.05$. Different letters (parentheses) in the same column indicate significant differences, $P < 0.05$.

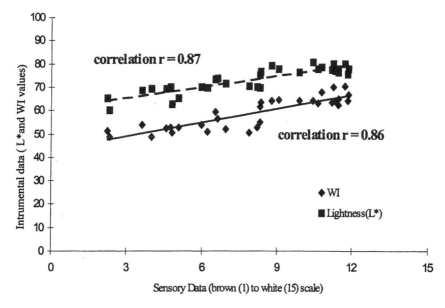

Figure 16.1 Correlation coefficient determination between sensory and instrumental analysis to evaluate color change in apple slices.

study demonstrates that the instrumental technique can be just such an effective measure of the browning changes in apple slices.

CONCLUSIONS

Instrumental values of apple slices' lightness (L^*) and WI were found to have a correlation coefficient of $r = 0.87$ with respect to sensory data. The enzymatic browning of the apple slices was significantly ($P < 0.05$) inhibited by polysaccharide film for 12 days at both refrigeration and room temperatures. Browning was also significantly ($P < 0.05$) inhibited by storage at refrigeration temperatures; however, this effect was significantly ($P < 0.05$) lower than that of the polysaccharide film.

REFERENCES

Avena-Bustillos, R. J., Cisneros Zeballos, L. A., Krochta, J. M., and Saltveit, M. E. 1993. Optimization of edible coatings on minimally processed carrots using response surface methodology. *ASAE* 36 (3):801–805.

Baldwin, E. A., Nisperos-Carriedo, M. O., and Baker, R. A. 1995. Edible coatings for the lightly processed fruits and vegetables. *Hortscience.* 30(1):35–37.

Bartolome, A. P., Ruperez, P., and Fuster, C. 1996. Freezing rate and frozen storage effects on color and sensory characteristics of pineapple fruit slices. *J. Food Sci.* 61 (1):154–156.

Guilbert, S. 1986. *Technology and Application of Edible Protective Film in Food Packaging.* London: Elsevier Applied Science.

Judd, D. B., and Wyszecki, G. 1963. Tools and techniques. In *Color in Business, Science, and Industry,* 2nd Ed., pp. 298–301. New York: John Wiley and Sons, Inc.

Kader, A. A., Stevens, M. A., Aalbright-Holton, M., Morris, L. L., and Algazi, M. 1977. Effect of fruit ripeness when picked on flavor and composition of fresh market tomatoes. *J. Amer. Soc. Hort. Sci.* 102:724.

Kinzel, B. 1992. Protein-rich edible coatings for food. *Agr. Res.* May: 20–21.

Kostyla, A. S., and Clysdesdale, F. M. 1978. The psychophysical relationship between color and flavor. *Crit. Rev. Food Sci. Nutr.* 10:310–321.

Nicolas, J. J., Richard-Forget, F. C., Goupy, P. M., and Aubert, S. Y. 1994. Enzymatic browning reactions in apple and apple products. *Crit. Rev. Food Sci. Nutr.* 34 (2):109–157.

Pennisi, E. 1992. Sealed in edible film. *Sci. News* 141:12.

Resurreccion, A. V. A., and Shewfelt, R. L. 1985. Relationships between sensory attributes and objective measurements of postharvest quality of tomatoes. *J. Food Sci.* 50:1242–1245.

Shewfelt, R. L. 1987. Quality of minimally processed fruits and vegetables. *J. Food Quality* 10:143–156.

Uhlemann, O. 1993. Controlling the browning of apple slices through coating with carbohydrates. In *Report to the Washington Tree Fruit Research Association.* USDA Western Regional Research Center, 800 Buchanan St. Albany, CA 94710.

Wong, D. W. S., Tillin, S. J., Hudson, J. S., and Pavlath, A. E. 1994. Gas exchange in cut apples with bilayer coatings. *J. Agric. Food Chem.* 42:2278–2285.

CHAPTER 17

Effect of Polysaccharide Film on Ethylene Production and Enzymatic Browning of Apple Slices

NICOLÁS BRÁNCOLI
GUSTAVO V. BARBOSA-CÁNOVAS

INTRODUCTION

MINIMALLY processed fruits and vegetables are able to maintain their fresh-like state from the study of their physiology, which involves wounded tissue. Minimal types of processing, such as abrasion, peeling, slicing, chopping, or shredding, differ from traditional thermal processing in that the tissue remains viable or "fresh" during subsequent handling (Rolle and Chism, 1987; Miller, 1992; King and Bolin, 1989). The behavior of fresh fruits and vegetable tissue includes increased respiration, ethylene production, and other chemical and physical effects, such as oxidative browning reactions, lipid oxidation, and water loss. Therefore, minimizing these deteriorative effects will result in increased shelf life and greater maintenance of appearance, flavor, and nutritional quality (Brecht, 1995).

Various approaches are used to minimize the deteriorative effects caused by minimal processing. Among these are low-temperature storage, special preparation techniques, use of additives, modified/controlled atmosphere, and application of edible films (King and Bolin, 1989; Cantwell, 1992). Edible films also offer a possible method of extending the shelf life of minimally processed commodities. When produce is coated with edible films, a semipermeable barrier to gases and water vapor is obtained. Edible films can reduce respiration and water loss as well as act as carriers of preservatives and antioxidants (Kester and Fennema, 1986; Guilbert, 1986; Krochta, 1991, 1992). Most investigations of edible films focus primarily on the water migration across the film. Little attention is directed to the gas transport and suppression of undesirable physiological changes that may well determine the success of

233

using edible coatings on minimally processed fruits and vegetables (Wong et al., 1994). Therefore, the objective of this study was to investigate the effect of using different polysaccharide film formulations to prevent browning reactions and to decrease ethylene production in apples.

MATERIALS AND METHODS

APPLES

Red Delicious apples were obtained from a local supermarket and stored at 4°C until analyzed. For color evaluation, the apples were washed, sliced for uniform size, and cored. For ethylene analysis, the apples were washed and cut into cylinders of 1.6 cm in diameter × 2.5 cm.

EDIBLE COATINGS

Both apple cylinders and slices were dipped in a solution containing ascorbic acid, calcium chloride, glycerol, potassium sorbate, maltodextrin, and methylcellulose. All chemicals were reagent grades and were obtained from SIGMA® (potassium sorbate, calcium chloride), Spectrum® (ascorbic acid), DOW® (methylcellulose), GPC® (maltodextrin), and Aldrich® (glycerol).

To assay the effect of polysaccharide concentration on ethylene production and color change of apple pieces, eight formulas were selected. Each formula contained a common base of ingredients consisting of 1% ascorbic acid, 1% glycerol, 0.1% potassium sorbate, and 0.25% calcium chloride. The selected polysaccharide contents for the eight formulas were

Formula	Maltodextrin [% (w/w)]	Methylcellulose [% (w/w)]
1	0.5	0.0
2	1.0	0.0
3	1.5	0.0
4	0.0	0.5
5	0.0	1.0
6	0.0	1.5
7	0.5	1.0
8	1.0	0.5

The effect of the plasticizer on ethylene production and color change in apple pieces was determined with film formulations containing 0%, 1%, 2%, and 5% glycerol. The common base of each formula was 1% ascorbic acid, 0.1% potassium sorbate, 0.25% calcium chloride, and 1.5% methylcellulose.

To evaluate the effect of calcium chloride on ethylene production and color change of apple pieces, 0%, 0.25%, and 0.5% calcium chloride concentrations were used. For this case, the common base was 1% ascorbic acid, 1% glycerol, 0.1% potassium sorbate, and 1.5% methylcellulose.

ETHYLENE EVALUATION

The effect of the different edible film formulations on ethylene evolution was determined by analyzing the headspace gas composition. Three apple cylinders, coated with a given treatment, were stored in a 270-ml sealed glass container at $20 \pm 1°C$. Headspace samples were withdrawn every 8 hours and analyzed for ethylene (C_2H_4) by gas chromatography (GC). The GC used for ethylene analyses was a Packard® model 427 GC equipped with a flame ionization detector. The column consisted of a 2 m × 0.3 cm stainless-steel column packed with Poropak Q. The column temperature was 50°C, while the injector and detector temperatures were 100°C. The ethylene production rate was calculated by regression analysis of the linear slope of the curve before saturation of ethylene in the glass container. The percent of ethylene production rate reduction was calculated by comparing the ethylene production rate of the coated apple cylinder with the respective control, or uncoated apple cylinder (Wong et al., 1994).

COLOR EVALUATION

The effect of the different edible film formulations on browning discoloration was determined by analyzing the color evolution of the apple slices through time. The uncoated and coated apple slices were thus packaged separately in plastic bags, and color lightness (L^*) was determined using a Minolta® colorimeter model 2002 (Minolta, Camera Co. Ltd., Japan) after 1, 2, 3, 4, and 5 days in storage at 4°C. The Minolta® colorimeter determines color based on the Hunter modified L^*, a^*, b^* (CIE) system where $L^* = 100$ is white, $L^* = 0$ is black, $a^* > 0$ is red, $a^* < 0$ is green, $b^* > 0$ is yellow, and $b^* < 0$ is blue. L^* measurements were plotted against time. Regression analysis was used to determine the best-fit curve for the data points. The percentage of browning inhibition was determined for all the edible film treatments to allow their comparison according to the following formula (similar to the one used by Uhlemann in 1993):

$$Inh \% = \frac{(L^*_{0,0} - L^*_{0,f}) - (L^*_0 - L^*_f)}{(L^*_{0,0} - L^*_{0,f})} \tag{1}$$

where
$L^*_{0,0}$ = initial lightness value of uncoated apple slices
L^*_{0f} = final lightness value of uncoated apple slices
L^*_0 = initial lightness value of coated apple slices
L^*_f = final lightness value of coated apple slices

STATISTICAL ANALYSIS

Differences in the ethylene and color evaluation parameter were statistically analyzed using one-factor analysis of variance (ANOVA) and were classified by Fisher's LSD test ($P < 0.05$) employing SAS computer software.

RESULTS AND DISCUSSION

POLYSACCHARIDE CONCENTRATION OF EDIBLE FILMS

Polysaccharide edible film treatments were found to retard surface discoloration in the studied apple slices. Select polysaccharide edible film formulas reduced the rate of surface discoloration relative to the uncoated apple slices. This effect was expressed by a high inhibition percentage (Inh. %).

Each polysaccharide formula (Table 17.1) demonstrated that, as the polysaccharide concentration increased, the inhibition percentages of surface discoloration on the apple slices (Inh. %) also increased significantly. Thus, the presence of a larger polysaccharide concentration leads to a lower rate of discoloration in apple slices. There were no significant differences in the inhibition parameter between the methylcellulose and maltodextrin polysaccharides selected. Indeed, 1.5% (weight/weight) maltodextrin and 1.5% (weight/weight) methylcellulose film formulations showed the best surface discoloration and rate of decreasing ethylene production. When both polysaccharides were used at the same time in the edible film formulation at a total level of 1.5% (weight/weight) polysaccharide concentration, the discoloration was more noticeable than in the two above mentioned edible film formulations.

In addition, the use of a large concentration of maltodextrin and methylcellulose, within the edible film formulation, decreased the ethylene production rate of apple cylinders even more, noted in Table 17.2 as the ethylene rate reduction percentage between the coated and uncoated control. This ethylene rate production suppression effect was similar to the one obtained by Park et al. (1994) for the content of polysaccharide contained within the film, which decreased gas permeability as the content of the polymer was increased. Ethylene rate production suppression effect is also supported by McHugh et

TABLE 17.1. Inhibition of Color Change of Apple Slices Using Different Edible Film Formulas.*

Polysaccharide	
Formula	Inh. %
0.5% Methylcellulose	84.77[a]
1.0% Methylcellulose	90.77[b]
1.5% Methylcellulose	95.27[c]
0.5% Maltodextrin	90.83[b]
1.0% Maltodextrin	91.88[b]
1.5% Maltodextrin	95.95[c]
0.5% Methylcellulose + 1.0% Maltodextrin	91.14[b]
1.0% Methylcellulose + 0.5% Maltodextrin	93.20[d]
Calcium Chloride	
Formula	
0.00% Calcium Chloride	84.18[a]
0.25% Calcium Chloride	94.53[b]
0.50% Calcium Chloride	95.04[c]
Glycerol	
Formula	
0.0% Glycerol	83.11[a]
1.0% Glycerol	89.03[b]
2.0% Glycerol	89.34[b]
5.0% Glycerol	59.56[c]

* Means of three replications of 20 measurements. Different letters in the same column indicate significant differences, $P < 0.05$.

al. (1993), Park and Chinnan (1993), and Pascat (1986) who attributed it to the increase in film thickness as the concentration of solids within the film is increased. This study found no significant differences in the reduction of ethylene production rates between the two polysaccharides used. Therefore, the decrease in ethylene production rate seems to support the fact that the polysaccharide concentration within films significantly ($P < 0.05$) affects oxygen permeability. In fact, both ethylene production and enzymatic browning depend on the presence of oxygen to develop a given reaction (Brecht, 1995).

The use of surface discoloration as an optimization method is not new. Avena-Bustillos et al. (1993) used the surface discoloration assay with peeled carrots to optimize the sodium caseinate/stearic acid ratio influence. Uhlemann (1993) used a similar technique to optimize the alginate/carrageenan ratio within a film applied to apple slices. In addition, the ethylene production technique seems to be an effective technique to indirectly determine the oxygen suppression used previously on apple pieces by Wong et al. (1994).

TABLE 17.2. Observed Evolved Ethylene Rate Reduction in the Headspace
of Coated and Uncoated Apple Pieces Using Different Edible Film
Formulations.*

Polysaccharide	
Formula	Ethylene Rate Reduction (%)
0.5% Methylcellulose	47.9 ± 15.6[a]
1.0% Methylcellulose	57.1 ± 9.59[b]
1.5% Methylcellulose	62.9 ± 10.48[c]
0.5% Maltodextrin	32.3 ± 9.95[d]
1.0% Maltodextrin	55.5 ± 9.20[b]
1.5% Maltodextrin	61.7 ± 5.69[c]
0.5% Methylcellulose + 1.0% Maltodextrin	57.4 ± 4.02[b]
1.0% Methylcellulose + 0.5% Maltodextrin	53.2 ± 2.83[e]
Calcium Chloride	
Formula	
0.00% Calcium Chloride	25.3 ± 3.15[a]
0.25% Calcium Chloride	52.9 ± 3.27[b]
0.50% Calcium Chloride	64.5 ± 3.22[c]
Glycerol	
Formula	
0.0% Glycerol	32.1 ± 14.94[a]
1.0% Glycerol	54.0 ± 3.80[b]
2.0% Glycerol	52.7 ± 4.32[b]
5.0% Glycerol	30.2 ± 1.59[a]

* Means of eight experimental sets ± SD. Each experimental set consists of one coated sample and
its control, both taken from the same apple. Different letters in the same column indicate significant
differences, $P < 0.05$.

GLYCEROL-PLASTICIZER CONCENTRATION OF EDIBLE FILMS

Plasticizers reduce internal hydrogen bonding and increase intermolecular
spacing, thereby decreasing brittleness and increasing the permeability of film
materials (Lieberman and Gilbert, 1973). The effects of glycerol plasticizer
on ethylene production rate and apple tissue discoloration observed in this
study are presented in Tables 17.1 and 17.2. The reduction of ethylene production rates significantly ($P < 0.05$) increased from 0% to 1% glycerol concentration, but there was no significant difference in the ethylene production rates
between the 1% and 2% glycerol concentration. However, at 5% glycerol
concentration, the reduction in the rate of ethylene production at 30.2% was
not significantly different from the 32.1% obtained when glycerol was not
present in the edible film formula.

The inhibited surface discoloration was similar at 1% and 2% glycerol
content, but was significantly different from that obtained at 0% and 5%

glycerol concentration. Ethylene and surface discoloration thus indirectly showed an increase in the oxygen permeability on the apple tissue. Researchers reported that this increase in oxygen permeability due to the presence of a plasticizer may be dependent on the concentration and type of plasticizer and base polysaccharide (Banker et al., 1966; Park et al., 1992; Park and Chinnan, 1993).

CALCIUM CHLORIDE CONCENTRATION OF EDIBLE FILMS

The addition of calcium chloride to edible film formulations is not well studied, though it is known that calcium chloride may be incorporated into edible films to improve the texture and color of food products. Certain polysaccharide edible films, such as alginate and low methoxyl pectin, require solutions of calcium chloride to induce gelation of films (Swenson et al., 1953; Earle, 1968). Calcium chloride also influences cellular respiration, affecting enzyme or metabolite reactions such as the ethylene production rate (Rosen and Kader, 1989). Calcium chloride minimizes apple tissue softening at concentrations greater than 0.3% due to the cross-linking of pectins. However, concentrations of calcium chloride larger than 0.3% present appearance and flavor problems in apple slices (Monsalve-González et al., 1993).

This research revealed that ethylene production was significantly decreased by the addition of calcium chloride to the film formulation. Table 17.2 shows that as the calcium concentration was increased from 0% to 0.25%, the percent of ethylene rate reduction was increased by almost 100%. Similar results were obtained by Wong et al. (1994) for the reduction of ethylene production rates using bilayer edible film formulations.

The inhibition of surface discoloration (Inh. %) presented less significant changes than those observed in the ethylene production rate in films with and without calcium chloride. Indeed, Table 17.1 shows a change from the film using no calcium chloride to the film with 0.25% calcium chloride. Also, no significant difference in the percent of surface discoloration was found between the films that contained 0.25% and 0.50% calcium chloride film. Therefore, increasing the calcium chloride concentration is not an appropriate solution to avoiding problems such as surface browning or ethylene production because of the possible undesirable bitter flavor and appearance produced in apple pieces.

CONCLUSIONS

The rates of ethylene production and surface discoloration were shown to decrease using polysaccharide film formulations. As the polysaccharide percentage within the film was increased, both ethylene production rate and surface browning of the apple slices treated with the film were decreased. Both 1.5% (weight/weight) maltodextrin and 1.5% (weight/weight) methylcellulose film formulations showed the best discoloration and ethylene rate reduction. The addition of glycerol to the polysaccharide film solution in concentrations greater than 1% or 2% (weight/weight) enhanced the surface discoloration of the apple pieces. Therefore, the best results were obtained when the formulation consisted of 1.5% polysaccharide, 1% ascorbic acid, 1% glycerol, 0.1% potassium sorbate, and 0.25% calcium chloride. These assays of ethylene production rate and surface discoloration may lead to further applications in optimizing formulations of edible film.

REFERENCES

Avena-Bustillos, R. J., Cisneros-Zeballos, L. A., Krochta, J. M., and Saltveit, M. E. 1993. Optimization of edible coatings on minimally processed carrots using response surface methodology. *ASAE.* 36 (3):801–805.

Banker, G. S., Gore, A. Y., and Swarbrick, J. 1966. Water vapor transmission properties of free polymer films. *J. Pharm. Pharmac.* 18:457.

Brecht, J. K. 1995. Physiology of lightly processed fruits and vegetables. *Hort. Sci.* 30 (1):18–22.

Cantwell, M. 1992. Postharvest handling systems: Minimally processed fruits and vegetables. In *Postharvest Technology of Horticultural Crops.* 2nd ed. University of California, Division of Agriculture Natural Resources, Oakland.

Earle, R. D. 1968. US Patent 3,395,024.

Guilbert, S. 1986. *Technology and Application of Edible Protective Film in Food Packaging.* London: Elsevier Applied Sci.

Kester, J. J., and Fennema, O. R. 1986. Edible films and coatings: A review. *Food Technol.* 12:47–59.

King, A. D., and Bolin, H. R. 1989. Physiological and microbial storage stability of minimally processed fruit and vegetables. *Food Technol.* 43:132–139.

Krotcha, J. M. 1991. Innovations for tomorrow's foods. *USA Today.* 1:78–79.

Krotcha, J. M. 1992. Control of mass transfer in foods with edible coatings and films. In *Advances in Food Engineering.* R. P. Singh and M. A. Wirakartakusumah, eds. New York: CRC Press.

Lieberman, E. R., and Gilbert, S. G. 1973. Gas permeation of collagen films as affected by cross-linkage, moisture, and plasticizer content. *J. Polymer Sci.* 41:33–43.

McHugh, T. H., Avena-Bustillos, R. J., and Krotcha, J. M. 1993. Hydrophilic edible films. Modified procedure for water vapor permeability and explanation of thickness effects. *J. Food Sci.* 58 (4):899.

Miller, A. R. 1992. Physiology, biochemistry and detection of bruising (mechanical stress) in fruits and vegetables. *Postharv. News Info.* 3:53–58.

Monsalve-González, A., Barbosa-Cánovas, G. V., and Cavalieri, R. P. 1993. Mass transfer and textural changes during processing of apples by combined methods. *J. Food Sci.* 58 (5):1118–1124.

Park, H. J., and Chinnan, M. S. 1993. Gas and water vapor barrier properties of edible coatings for fruits and vegetables. *J. Food Eng.* 20:352–359.

Park, H. J., Weller, C. L., Vergano, P. J., and Testin, R. F. 1992. Factors affecting barrier and mechanical properties of protein-based edible, degradable films. Paper No. 428, presented at the *52nd Annual Meeting of the Institute of Food Technologists,* New Orleans, LA, June 20–24.

Park, H. J., Bunn, J. M., Vergano, P. J., and Testin, R. F. 1994. Gas permeation and thickness of the sucrose polyesters, semperfresh coatings on apples. *J. Food Proc. Pres.* 18:349–358.

Pascat, B. 1986. Study of some factors affecting permeability. In *Food Packaging and Preservation. Theory and Practice,* M. Mathlouthi, ed., London: Elsevier Applied Science.

Rolle, R., and Chism, G. W. 1987. Physiological consequences of minimally processed fruits and vegetables. *J. Food Qual.* 10:157–165.

Rosen, J. C., and Kader, A. A. 1989. Postharvest physiology and quality maintenance of sliced pear and strawberry fruits. *J. Food Sci.* 54:656–659.

Swenson, H. A., Miers, J. C., Schultz, T. H., and Owens, H. S. 1953. Pectinate and pectate coatings. II. Application to nut and fruit products. *Food Technol.* 7:232–235.

Uhlemann, O. 1993. Controlling the browning of apple slices through coating with carbohydrates. In *Report to the Washington Tree Fruit Research Association.* USDA Western Regional Research Center, St. Albany, CA.

Wong, D. W. S., Tillin, S. J., Hudson, J. S., and Pavlath, A. E. 1994. Gas exchange in cut apples with bilayer coatings. *J. Agric. Food Chem.* 42:2278–2285.

CHAPTER 18

Quality Changes during Refrigerated Storage of Packaged Apple Slices Treated with Polysaccharide Films

NICOLÁS BRÁNCOLI
GUSTAVO V. BARBOSA-CÁNOVAS

INTRODUCTION

CONSUMER demand for fresh fruits and vegetables, as well as convenience and safety of consumption, is promoting interest in a minimally processed product with fresh-like qualities. Such products vary widely according to the characteristics of the unprocessed commodities and forms for consumption. Minimally processed foods are usually raw tissues that remain viable after minimal processing. Physiological and biochemical changes in minimally processed foods may occur at faster rates than in intact raw commodities due to the tissue damage (Brecht, 1995; Huxsoll et al., 1989; Watada et al., 1990; Shewfelt, 1987).

To retard or prevent quality loss in minimally processed foods, various treatments can be applied. Specifically, research on minimally processed apple slices is focused on non-sulfite methods to inhibit browning due to the ban on the use of sulfites (FDA, 1987). Various alternatives to the use of sulfites including reducing agents, acidulants, chelating agents, inorganic salts, and enzymes are being evaluated for use on fresh cut apples (Macheix et al., 1990). One possible alternative for preserving fresh apple slices is edible film technology. Apple slices can be coated in edible films, providing a semipermeable barrier to gases and water vapor. Edible films can reduce respiration and water loss and can be used as carriers of preservatives and antioxidants (Baldwin et al., 1995).

Several types of edible films are being used for preservation of minimally processed products. Mixtures of sucrose fatty acid esters are being used for coating fresh fruits and vegetables (Banks, 1984; Santerre et al., 1989). Soy-

243

bean edible film developed at the USDA preserved the freshness of apple slices in preliminary studies (Kinzel, 1992). Browning and water loss of cut apple slices is inhibited by a chitosan and lauric acid composite edible film (Pennisi, 1992). Most investigations focused on the effects of edible films on moisture retention. There is relatively little attention directed to other factors, such as gas permeability, physical appearance, and physiological changes of minimally processed fruits and vegetables due to the application of edible films. Indeed, modification of internal gas composition following the application of edible films can increase disorders due to high CO_2 and low O_2 (Smith et al., 1987). Several problems are associated with edible films, including anaerobic fermentation of apples and bananas, high levels of core flush in the apple, and rapid weight loss in tomatoes (Banks, 1984; Smith and Stow, 1984). Therefore, the effectiveness of edible films for minimally processed fruits and vegetables depends primarily on the selection of appropriate films, which results in physiological, biochemical, physical, and microbial stability (Kim et al., 1993). The specific objectives of this research were to coat apple slices with a maltodextrin edible film and determine the effects on ethanol content, soluble solids, titratable acidity, color, firmness, weight loss, and microbial populations of the fruit during storage.

MATERIALS AND METHODS

FRUIT

Red Delicious apples were obtained from a local supermarket and were kept in cold storage until the experiment. A total of 270 apples were divided into three replications. Each lot of 90 apples was subdivided into three groups, two different coated apple slice treatments and one control or untreated one.

POLYSACCHARIDE FILM

Apple slices were dipped in a solution containing ascorbic acid, calcium chloride, glycerol, potassium sorbate, and a polysaccharide (maltodextrin). All chemicals were reagent grade and were obtained from SIGMA® (potassium sorbate, calcium chloride), Spectrum® (ascorbic acid), GPC® (maltodextrin), and Aldrich® (glycerol).

To assess the effect of polysaccharide concentration of edible films on the quality changes of apple slices, two different formulas were used. Each formula contained a common base of ingredients consisting of 1% ascorbic acid, 1% glycerol, 0.1% potassium sorbate, and 0.25% calcium chloride. The polysaccharide contents of the two formulas were

Formula	Maltodextrin [% (w/w)]
1	0.5
2	1.5

After treatment with the edible film solution, apple slices were packaged in sealed polyethylene plastic bags and kept in cold storage at 4°C for 21 days.

ETHANOL CONTENT, SOLUBLE SOLIDS, AND TITRATABLE ACIDITY

Every two days, one apple slice sample from each treatment was analyzed for ethanol content. Typically, apple slices were blended to apple juice. A 1-ml sample of apple juice from each selected sample was prepared for ethanol analysis by the addition of 100 μl of 1 N trichloroacetic acid (TCA) and 10 mg of polyvinylpolypyrrolidone (PVPP) in a microcentrifuge tube. After centrifugation, 1 μg of supernatant was injected into a Packard® gas chromatograph (model 427) equipped with an 80/100 Poropack T column. The column temperature was 120°C. The injector and detector temperatures were 150°C. The ethanol content was expressed in ppm.

Five milliliters of juice from each apple were titrated with KOH to an endpoint of pH 8.1 with a Metohm® titroprocessor (model 682). The percentage of titratable acidity was calculated as grams of malic acid per milliliter of apple juice, usually expressed as percent of malic acid. The soluble solids of each treatment was determined with a Reichert® digital refractometer (model ABBE MARK II) and was expressed as °B. Ethanol, titratable acidity, and soluble solid determinations were made after 0, 3, 6, 9, 12, 15, and 21 days in storage at 4°C.

COLOR MEASUREMENT

The effects of edible film formulas on browning discoloration of coated apple slices were determined by analyzing the color of the apple slices. For measurement of flesh color, three apple slice samples from each treatment were selected. The flesh portion of each slice was exposed to a Minolta® colorimeter (model 2002) after 0, 3, 6, 9, 12, 15, and 21 days in storage at 4°C. The Minolta® colorimeter determines color lightness based on the Hunter modified L^*, a^*, b^* (CIE) system where $L^* = 100$ is white, $L^* = 0$ is black, $a^* > 0$ is red, $a^* < 0$ is green, $b^* > 0$ is yellow, and $b^* < 0$ is blue. Lightness $(L)^*$ measurements were obtained using the average of 20 measurements of each apple slice.

TEXTURE

To determine the effect of the edible film formulas on textural changes of the apple slices, cylinders of 10 mm in length and 13 mm in diameter were cut from eight apple slices of each treatment. The firmness of apples was determined with an Instron® Universal Testing machine (model 1350) using the uniaxial compression test at 0, 3, 6, 9, 12, 15, and 21 days. The probe used for determining the firmness was a flat plate applied to 75% compression. The compression force recorded was the peak force in the force-time curve output. The setting parameters for the probe were 3.0 mm/s for the pre-test speed, 1.0 mm/s for the test speed, and 3.0 mm/s for the post-test speed.

MICROBIAL ANALYSIS

Three 25-g apple slices were used from each edible film treatment to determine microbial counts at 0, 3, 6, 9, 12, 15, and 21 days. Total plate, mold, and yeast counts were determined as surface counts in 25 g of apple mixed with 225 ml of 0.1% sterile peptone dilution water. Total plate counts (APC) were determined in triplicate on decimal dilutions using Standard Plate Count Agar (DIFCO) and were incubated for 2 days at 35°C. Mold and yeast counts were determined using Potato Dextrose Agar (DIFCO) plus 10% w/w tartaric acid and were incubated for 5 days at 22°C.

WEIGHT LOSS

Twenty apple slices from each treatment were placed in a polyethylene plastic bag in a cold room at 4°C. The weights of the 20 apples were recorded at 0, 3, 6, 9, 12, 15, and 21 days of storage. The percentage of weight loss per initial fruit weight for each sample was calculated in triplicate, and the average was taken.

STATISTICAL ANALYSIS

Differences in quality characteristics on the application of edible film treatment on apple slices were statistically analyzed by one-factor ANOVA and were classified by Fisher's LSD test ($P < 0.05$). Statistical analyses employed SAS computer software.

RESULTS AND DISCUSSION

ETHANOL PRODUCTION

Ethanol is one of the principal volatiles produced by ripening apples. Ethanol typically accumulates in overripe and senescent apples (Fidler et al., 1973).

Ethanol accumulation is a product of anaerobic fermentation taking place in some fruits and vegetables (Smith et al., 1987). Apple tissue metabolizes ethanol as a result of exposure to low oxygen atmospheres (Fidler, 1968). Significant ($P < 0.05$) differences existed in ethanol content among the treatments in this experiment (Table 18.1). As the maltodextrin concentration increased from 0.5 to 1.5%, the ethanol content also significantly ($P < 0.05$) increased from days 9 and 21. No significant ($P < 0.05$) differences were observed prior to day 9. Also, considering the level of ethanol concentration after day 3, the ethanol content between both polysaccharide treatments and the control was significantly greater. Accumulation of ethanol and alcoholic off flavors are reported when the internal atmosphere of fruits and vegetables is affected by restricted gas exchange. Extremely low oxygen contents for broccoli and cauliflower resulted in off-flavors, as these products changed from aerobic to anaerobic metabolism. Also, tomatoes coated with corn-zein edible film produced ethanol during storage (Park et al., 1992).

FIRMNESS

Firmness values of uncoated and coated apple slices (Table 18.1) suggested that the maltodextrin edible film delayed softening. The firmness values of both coated apple slice treatments was significantly ($P < 0.05$) different from those obtained for the uncoated apple slice treatment or control. This difference in firmness was especially marked after day 6. There were no significant differences in the firmness values within each treatment except on days 0 and 21. Reduction in respiration rates of coated apple slices could be responsible for ripening delay, which resulted in reduction of softening during storage. Park et al. (1992) reported differences of 10% in firmness in coated tomatoes with lower respiration and oxygen consumption than the uncoated tomatoes.

Another cause of this delay in softening could be the inclusion of calcium within the formulation of the edible film. The role of calcium in plant tissues is not completely clear, although calcium maintains cell wall structures in fruits and vegetables (Poovaiah, 1986; Evensen, 1984; Morris et al., 1985; Rosen and Kader, 1989; Monsalve-González et al., 1993).

COLOR CHANGE

Maltodextrin edible film treatments retarded the surface discoloration of apple slices. Table 18.1 shows that the two edible film-coated apple slice treatments significantly ($P < 0.05$) reduced the rates of surface discoloration relative to non-coated apple slices or control. Apple slices treated with maltodextrin edible films presented higher lightness values than non-coated apple slices. Also, retardation in surface discoloration was affected by the maltodextrin concentration within the film. Lightness values after day 9 were signifi-

TABLE 18.1. Physical/Chemical Attributes of Apple Slices Treated with Maltodextrin Films during Storage at 4°C.*

	Treatment (% Maltodextrin Concentration in Film)		
	0.5	1.5	Control
Storage day	Firmness (compression force (N))		
0	57200[a(a)]	54235[a(b)]	55264[a(b)]
3	58920[a(a)]	55487[a(b)]	50157[b(c)]
6	56231[a(a)]	52468[c(b)]	50234[b(c)]
9	52268[b(a)]	57687[a(b)]	48265[c(c)]
12	54781[b(a)]	59842[b(b)]	48784[c(c)]
15	56457[a(a)]	54267[a(b)]	40634[d(c)]
21	48236[c(a)]	51234[c(b)]	43248[e(c)]
	Color (lightness value "L" value)		
0	81.2[a(a)]	81.5[a(a)]	80.9[a(a)]
3	79.3[a(a)]	80.5[a(a)]	71.4[b(b)]
6	78.2[b(a)]	78.5[b(a)]	69.2[c(b)]
9	77.6[b(a)]	78.1[b(a)]	67.5[d(b)]
12	75.9[c(a)]	77.2[b(b)]	64.5[e(c)]
15	73.6[d(a)]	76.5[c(b)]	62.2[f(c)]
21	69.4[e(a)]	72.6[d(b)]	59.4[g(c)]
	Ethanol concentration (ppm)		
0	28[a(a)]	27[a(a)]	29[a(a)]
3	47[b(a)]	55[b(a)]	32[a(b)]
6	150[c(a)]	157[c(a)]	56[b(b)]
9	185[d(a)]	200[d(b)]	58[b(c)]
12	205[e(a)]	210[e(a)]	75[c(b)]
15	208[f(a)]	254[f(b)]	96[d(c)]
21	270[g(a)]	352[g(b)]	102[e(c)]
	Soluble solids (°Brix)		
0	12.2[a(a)]	12.2[a(a)]	12.3[a(a)]
3	12.2[a(a)]	12.2[a(a)]	12.2[a(a)]
6	12.3[a(a)]	12.2[a(a)]	12.2[a(a)]
9	12.2[a(a)]	12.3[a(a)]	12.2[a(a)]
12	12.4[a(a)]	12.2[a(a)]	12.4[a(a)]
15	12.2[a(a)]	12.3[a(a)]	12.4[a(a)]
21	12.3[a(a)]	12.4[a(a)]	12.3[a(a)]
	Titratable acid (%, as malic acid)		
0	0.340[a(a)]	0.352[a(b)]	0.348[a(b)]
3	0.335[a(a)]	0.342[b(a)]	0.342[a(a)]
6	0.328[b(a)]	0.339[b(b)]	0.315[b(c)]
9	0.328[b(a)]	0.334[b(a)]	0.300[b(b)]
12	0.327[b(a)]	0.336[c(b)]	0.298[c(c)]
15	0.316[c(a)]	0.321[d(a)]	0.287[d(b)]
21	0.310[c(a)]	0.315[d(a)]	0.290[e(b)]

*Means of triplicate measurements, except for color which is a mean of 16 measurements. Different letters in the same column indicate significant differences, $P < 0.05$. Different letters (parentheses) in the same row indicate significant differences, $P < 0.05$.

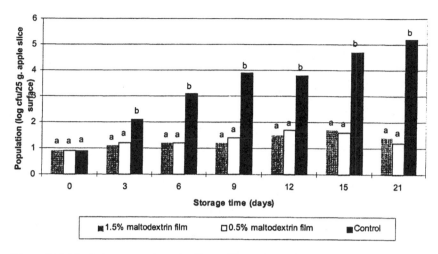

Figure 18.1 Total plate counts in apple slices (different letters indicate significant differences, $P < 0.05$).

cantly ($P < 0.05$) different between the two maltodextrin film-coated apple slice treatments. Apple slices treated with 1.5% maltodextrin concentration had the largest lightness value after 21 days of storage at 4°C.

Discoloration of minimally processed apple slices occurs at the cut surface as a result of the disruption of compartmentation of the cell structures. The disruption occurs when cells are broken, allowing substrates and oxidases to come in contact (Rolle and Chism, 1987). Browning at the cut surface is a very important limiting factor in the acceptance of many minimally processed fruits and vegetables (Brecht, 1995). Apple browning phenomenon is the result of three factors: polyphenoloxidase enzyme, polyphenolic substrates, and oxygen (Nicolas et al., 1994). As shown indirectly in the present study, the edible film used on apple slices acted as a barrier to oxygen, preventing the enzymatic browning reaction.

MICROBIAL POPULATION

Apple slices treated with the maltodextrin edible film had significantly ($P < 0.05$) lower populations of molds and yeasts (Figures 18.1 and 18.2). Mold and yeast counts showed significant differences of almost 1 log on day 3 to 3 log on day 21. In this case, the maximum mold and yeast levels reached by the apple slice treated with the edible film was 2.7 log, while the control reached 4.5 log. Populations of molds, yeasts, and aerobic bacteria were not significantly different among edible film-coated apple slice treatments. These results support the use of potassium sorbate as an antimicrobial source within the edible film formulation.

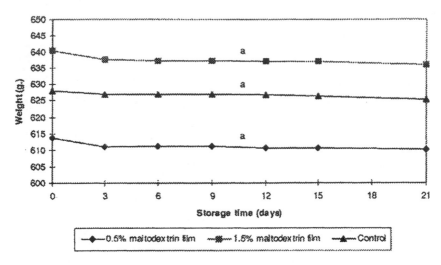

Figure 18.2 Population of molds and yeasts in apple slices (different letters indicate significant differences, $P < 0.05$).

The use of some antimicrobial agent is essential because of factors such as the high relative humidity imparted by the coating and promotion of an anaerobic environment. In the case of potassium sorbate as an antimicrobial, the primary inhibitory action of sorbates is against yeasts and molds; the activity against bacteria is not comprehensive and appears to be selective. Extensive research during the 1950s demonstrated the impressive effectiveness of sorbates against yeasts and molds and resulted in the extensive use of the compounds as fungistatic agents in many foods. Effective antimicrobial concentrations of sorbates in most foods are in the range 0.05–0.30% (Davidson and Branen, 1993).

Figures 18.1 and 18.2 show that the bacterial counts obtained (APC) were very low. This situation is due to factors such as the acidity of the fruit (pH 4.05) and the anaerobic environment. The pH range of apple slices represents an uncommonly acidic environment for the development of bacterial populations. Indeed, most of the microbial contamination in fruits is due to mold and yeast growth (Jay, 1992).

SOLUBLE SOLIDS, TITRATABLE ACIDITY, AND WEIGHT LOSS

The soluble solids content of all the apple slice treatments were not significantly ($P < 0.05$) different. No significant difference was found in any of the treatments that kept their soluble solids content of 12.2° Brix constant (Table 18.1). Titratable acidities of the apple slices showed significant differences among the treatments. Edible film-coated apple slices showed a higher acid content than non-coated apples, which was especially remarkable after day 9. The difference in acidity between coated and non-coated apple slices might be due to differences in the respiration behavior between the maltodextrin-coated apple slices and the non-coated apple slices. Acids are utilized quickly during respiration compared to other compounds (Moller and Palmer, 1984).

The weight loss of apple slices during the 21 days of storage was determined by comparing the initial and final weights (Figure 18.3). The weight loss was minimal—in the range of 1%. There were no significant differences among all of the treatments. Some sliced apple weight loss was expected because of the high transpiration that occurs through the peeled surface. Plant tissues are generally in equilibrium with the atmosphere at the same temperature and relative humidity of 99% to 99.5%. Any reduction of the water vapor pressure in the atmosphere below 99% in the tissue results in water loss. Avoiding desiccation at the cut surface of some minimally processed fruits and vegetables is critical for maintaining acceptable visual appearance (Brecht, 1995).

CONCLUSIONS

The maltodextrin film delayed surface discoloration and softening during the 21-day refrigerated storage study. The maltodextrin edible film caused ethanol production within the apple tissue. The weight loss was minimized using plastic bags. Titratable acidity values of apple slices coated with the maltodextrin film decreased during the 21 days. The maltodextrin film retarded

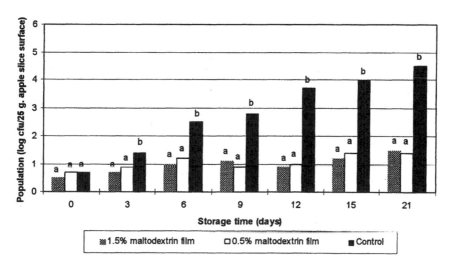

Figure 18.3 Weight variation in coated apple slices with time (different letters indicate significant differences, $P < 0.05$).

the growth of molds and yeasts on apple slices, thus extending their shelf life. However, further sensory studies are needed to validate the techniques explored in this work.

REFERENCES

Baldwin, E. A., Nisperos-Carriedo, M. O., and Baker, R. A. 1995. Edible coatings for the lightly processed fruits and vegetables. *Hort. Sci.* 30 (1):35–37.

Banks, N. H. 1984. Some effects of TAL Pro-long coating on ripening bananas. *J. Exp. Bot.* 35:127.

Brecht, J. K. 1995. Physiology of lightly processed fruits and vegetables. *Hort. Sci.* 30 (1):18–22.

Davidson, P. M., and Branen, A. L., eds. 1993. *Antimicrobials in Foods.* New York: Marcel Dekker, Inc.

Evensen, K. B. 1984. Calcium effects on ethylene and ethane production and 1-aminocyclopropane-1-carboxylic acid content in potato disks. *Physiol. Plant.* 60:125–128.

FDA. 1987. Chemical preservatives, Food and Drug Administration. *Code of Federal Regulation.* Title 21. Part 182; U.S.GPO, Washington, D.C.

Fidler, J. C. 1968. The metabolism of acetaldehyde by plant tissue. *J. Exp. Bot.* 19 (58):41–51.

Fidler, J. C., Wilkinson, B. G., Edney, K. L., and Sharples, R. O. 1973. *The Biology of Apple and Pear Storage.* Commonwealth Agr. Bur., Slough, U.K.

Huxsoll, C. C., Bolin, H. R., and King, A. D. 1989. Physiochemical changes and treatments for lightly processed fruits and vegetables. In *Quality Factors of Fruits and Vegetables—Chemistry and Technology.* Washington, DC: Amer. Chem. Soc.

Jay, J. M. 1992. *Modern Food Microbiology.* New York: Chapman & Hall.

Kader, A. A., Zagory, D., and Kerbel, E. L. 1989. Modified atmosphere packaging of fruits and vegetables. *CRC Crit. Rev. Food Sci. Nutr.* 28:1–30.

Kim, D. M., Smith, N. L., and Lee, C. Y. 1993. Quality of minimally processed apple slices from selected cultivars. *J. Food Sci.* 58 (5):1115.

Kinzel, B. 1992. Protein-rich edible coatings for food. *Agr. Res.* May: 20–21.

Macheix, J. J., Fleuriet, A., and Billot, J. 1990. *Fruit Phenolics.* Boca Raton, FL: CRC Press, Inc.

Moller, I. M., and Palmer, J. M. 1984. Regulation of the tricarboxylic acid cycle and organic metabolism. In *The Physiology and Biochemistry of Plant Respiration.* J. M. Palmer, ed., p. 105. London: Cambridge University Press.

Monsalve-González, A., Barbosa-Cánovas, G. V., and Cavalieri, R. P. 1993. Mass transfer and textural changes during processing of apples by combined methods. *J. Food Sci.* 58 (5):1118–1124.

Morris, J. R., Sistrunk, W. A., Sims, C. A., Main, G. L., and Wehunt, E. J. 1985. Effect of cultivar, postharvest storage, preprocessing dip treatments and style of pack on the processing quality of strawberries. *J. Amer. Soc. Hort. Sci.* 110:172.

Nicolas, J. J., Richard-Forget, F. C., Goupy, P. M., and Aubert, S. Y. 1994. Enzymatic browning reactions in apple and apple products. *CRC. Crit. Rev. Food Sci. Nutr.* 34 (2):109–157.

Park, H. J., Chinnan, M. S., and Shewfelt, R. L. 1992. Coating tomatoes with edible films: Prediction of internal oxygen concentration and effect on storage life and quality. Paper no. 848, presented at the *Annual Meeting of the Inst. of Food Technologists,* New Orleans, LA, June 20–24.

Pennisi, E. 1992. Sealed in edible film. *Sci. News.* 141:12.

Poovaiah, B. W. 1986. Role of calcium in prolonging storage life of fruits and vegetables. *Food Technol.* 40:86–89.

Rolle, R., and Chism, G. W. 1987. Physiological consequences of minimally processed fruits and vegetables. *J. Food Qual.* 10:157–165.

Rosen, J. C., and Kader, A. A. 1989. Postharvest physiology and quality maintenance of sliced pear and strawberry fruits. *J. Food Sci.* 54:656–659.

Santerre, C. R., Leach, T. F., and Cash, J. N. 1989. The influence of the sucrose polyester, Semperfresh, on the storage of michigam grown "McIntosh" and "Golden Delicious" apples. *J. Food Proc. Pres.* 13:293.

Shewfelt, R. L. 1987. Quality of minimally processed fruits and vegetables. *J. Food Qual.* 10:143–156.

Smith, S., Geeson, J., and Stow, J. 1987. Production of modified atmospheres in deciduous fruits by the use of films and coatings. *Hort. Sci.* 22 (5):772.

Smith, S. M., and Stow, J. R. 1984. The potential of a sucrose ester coating material for improving the storage and shelf life qualities of cox's orange pippin apples. *Ann. Appl. Biol.* 104:383.

Watada, A. E., Abe, K., and Yamauchi, N. 1990. Physiological activities of partially processed fruits and vegetables. *Food Technol.* 20:116, 117, 120–122.

Index

4-hexylresorcinol (4-HR) 74, 91
absorbance 45, 91, 153, 154
acceptability 111, 119, 120, 134, 135, 212, 213, 217, 220, 221
acetone 152
air 45, 75, 82, 142, 144, 147, 169, 180, 207, 210, 213, 214, 215, 216, 217, 218, 220
Alaska Pollock 103, 104, 106, 107, 108, 109, 112, 115, 116, 117, 118, 119, 120, 121
amino acids 57, 58, 69, 73, 74
amplitude sweep 59, 61, 64, 69
analysis of variance 105, 115, 213, 227, 236
antibrowning agent 90
antimicrobial 5, 128, 130, 138, 184, 225, 249, 250
antimicrobial agent 124, 128, 130, 183, 250
antioxidant 87, 89, 233, 243
apple 73, 74, 83, 87, 90, 91, 100, 128, 131, 132, 133, 134, 136, 140, 184, 185, 186, 187, 188, 189, 191, 193, 196, 197, 198, 199, 200, 201, 202, 203, 204, 205, 206, 209, 210, 211, 212, 213, 214, 216, 217, 220, 221, 222, 225, 226, 227, 229, 232, 234, 235, 236, 237, 238, 239, 240, 241, 243, 244, 245, 246, 251, 252, 253
apple slice 74, 75, 77, 78, 80, 81, 84, 85, 87, 91, 139, 185, 186, 187, 188, 191, 192,

193, 194, 195, 196, 198, 199, 200, 203, 205, 208, 209, 210, 211, 212, 213, 214, 215, 216, 217, 218, 221, 222, 225, 226, 227, 229, 231, 232, 235, 236, 237, 239, 240, 241, 243, 244, 245, 246, 247, 249, 250, 251, 252
ascorbic acid 73, 80, 84, 89, 91, 95, 100, 133, 135, 154, 160, 209, 214, 215, 222, 225, 226, 234, 235, 240, 244
Aspergillus flavus 129, 139
Aspergillus niger 44, 131, 139
Aspergillus ochraceus 129, 138, 139
asymptotic moduli 104, 105, 106, 108, 115
asymptotic modulus 104, 108, 113
atomic absorption 211
avocado 133, 134, 139, 150, 152, 153, 154, 156, 157, 159, 160

Bacillus coagulans 128, 129, 139
Bacillus subtilis 16, 30
bacterial spore 16, 17, 87, 165, 181
bacteriocin 5
banana 128, 131, 133, 139, 208, 244, 252
baroprotective effect 17, 44
biological 8, 14, 55, 109, 124, 126, 128, 131, 143, 163, 180, 210, 220
biscuits 144, 145